PUTTING TWO AND TWO TOGETHER

Selections from the Mathologer Files

PUTTING TWO AND TWO TOGETHER

Selections from the Mathologer Files

Burkard Polster
Marty Ross

Providence, Rhode Island

2020 *Mathematics Subject Classification.* Primary 00A08, 00A09, 00A67, 00A08, 97A80.

The AMS is pleased to publish these stories first seen in Australia's *The Age* newspaper. The views and opinions expressed herein are those of the authors and do not necessarily reflect the official position of the AMS.

For additional information and updates on this book, visit
www.ams.org/bookpages/mbk-141

Library of Congress Cataloging-in-Publication Data

Names: Polster, Burkard, author. | Ross, Marty, 1959– author.
Title: Putting two and two together : selections from the Mathologer files / Burkard Polster, Marty Ross.
Description: Providence, Rhode Island : American Mathematical Society, [2021] | Series: | "This is our second book of mathematics columns, following on from A Dingo Ate My Math Book."–Preface.
Identifiers: LCCN 2021036494 | ISBN 9781470460112 (paperback) | (ebook)
Subjects: LCSH: Mathematics–Popular works. | Mathematical recreations. | Mathematics–Study and teaching–Active programs. | AMS: General – General and miscellaneous specific topics – Recreational mathematics [See also 97A20]. | General – General and miscellaneous specific topics – Popularization of mathematics. | General – General and miscellaneous specific topics – Mathematics and architecture. | General – General and miscellaneous specific topics – Recreational mathematics [See also 97A20]. | Mathematics education – General, mathematics and education – Popularization of mathematics.
Classification: LCC QA93 .P654 2021 | DDC 510–dc23
LC record available at https://lccn.loc.gov/2021036494

∞ The paper used in this book is acid-free and falls within the guidelines
established to ensure permanence and durability.
Visit the AMS home page at `https://www.ams.org/`

10 9 8 7 6 5 4 3 2 1 26 25 24 23 22 21

For the little maths masters, Karl and Lara and Lillian and Eva,
and in loving memory of Marian Heckart Ross.

Contents

Preface

This is our second book of mathematics columns, following on from *A Dingo Ate My Math Book*.[1] It comprises another 1,000,000 *Maths Masters* columns, from the 11,111,111 columns we wrote in total. Only 1,111,111 to go.

Our columns appeared more or less weekly from 2007 to 2014, in Melbourne's *Age* newspaper. The mission of our columns, and of the book you are now holding, was to present ingenious, unusual and beautiful mathematical gems in as clear and as entertaining a manner as possible. These newspaper columns were pitched for the general reader, meaning we could assume little in mathematical background beyond a vague "Pythagoras" and a dusty x or two. Nonetheless, we were ambitious, and our intention was to not just present the gems to be viewed safely in their cabinets, but to *examine* them, to delve into the deep ideas. Whenever possible, our goal was – stone the crows! – to prove things.[2]

In 2009, *Maths Masters* column moved online, which allowed us much more freedom to explore ideas in depth, and to include many more illustrations. Our editors appeared puzzled at our willingness to sweat over *long* columns, still at a newspaper pay-rate, but somehow they never thought to stop us. This move online was the germination of Burkard's *Mathologer* YouTube channel, and this and our *Dingo* book can be considered a prehistory of *Mathologer*.

The columns as reproduced here are pretty much as they first appeared in *The Age*, except for updating, and other minor corrections and adjustments. Our first, *Dingo* book, contained articles with a particularly Australian theme or framing, but that is less the case here. Nonetheless, there are plenty of Australian references and Australianisms that are likely to be puzzling, and we have added many explanatory footnotes. We have also included a number of references to more technical articles in footnotes; when appropriate, we have indicated when such references are readily available online, but we have refrained from including precise but typically short-lived links. Finally, each column contains a (possibly tricky) puzzle or two, with solutions in an Appendix. We have also taken the opportunity to use some of these puzzles and solutions to tie up some technical loose ends, and to create some new ones.

This book is dedicated to our (now not so) little maths masters, Karl and Lara and Lillian and Eva. The book is also dedicated to the memory of Marian Heckart Ross, a passionate and unqualified teacher, and a passionate and very qualified mother and grandmother.

As for our previous book, we give our humble thanks to the wonderful Simon Pryor, CEO of the Mathematical Association of Victoria. Beyond Burkard himself, Simon is the single person most responsible for the birth of and form of *Mathologer*.

[1] American Mathematical Society, 2017.

[2] That's the way Aussies talk, even Honorary Aussies like us. You'll get used to it.

And thanks again, to Ken Merrigan our first editor at *The Age*, a kind and fiercely intelligent newspaper man from an era when newspapers were intelligent and had news. Thanks also to David Treeby, the third Amigo, for his very careful proof-reading. We would also like to thank the following people and organisations for their very kind permission to reproduce photographs in the book: Arthur Ganson, Henry Segerman and Stanford News Service. Thank you also to our friend and colleague John Zweck for his assistance with Chapter 20.

Finally, our thanks and our apologies to the beautiful, and (almost) always patient, Anu and Ying.

Credits

The American Mathematical Society gratefully acknowledges the kindness of these institutions and individuals in granting the following permissions:

Henry Segerman
Stereographic projection; see p. 52.

Creative Commons Attribution–Share Alike 2.0 generic license (`https://creativecommons.org/licenses/by-sa/2.0/deed.en`), by Laurent Errera from *L'Union*, France
Photograph of Malaysia Airlines Boeing 777-200ER (9M-MRO) taking off at Roissy–Charles de Gaulle Airport (LFPG) in France; see p. 81.

Creative Commons Attribution–Share Alike 3.0 Unported license (`https://creativecommons.org/licenses/by-sa/3.0/deed.en`), by Franz Richter
Photograph of Jared Tallent; see p. 115.

Creative Commons Attribution–Share Alike 4.0 International license (`https://creativecommons.org/licenses/by-sa/4.0/deed.en`).
Nick Riewoldt during the AFL round thirteen match between North Melbourne and St. Kilda on 16 June 2017 at Etihad Stadium in Melbourne, Victoria; see p. 107.
Portrait of Hermann Reichenau; see p. 221.

Stanford News Service
Photograph of Robert Osserman; see p. 183.

Frank R. Paul Estate
Cover of *Wonder Stories* (used with the acknowlegment of the estate); see p. 203.

Arthur Ganson
Machine and Concrete, photographs of the sculpture and a detail; see pp. 207–208.

Part 1

Putting Two and Two Together

CHAPTER 1

Cordial math

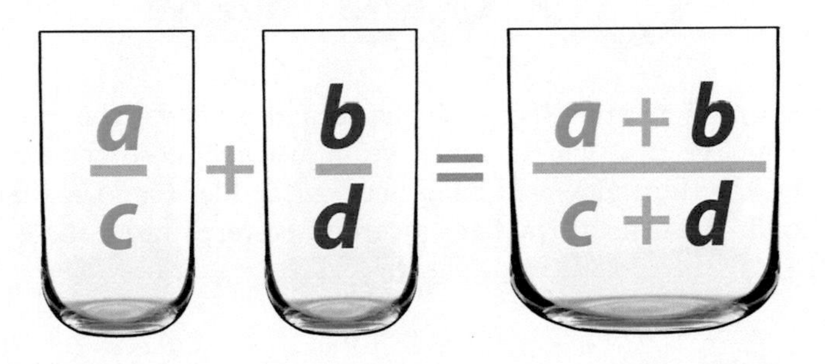

One of the trickiest topics in school mathematics is fractions. Why can't they just behave like familiar, friendly whole numbers? But your Maths Masters are here to help, and we've discovered a wonderful new way to add fractions. Here's an example:

$$\frac{4}{7} + \frac{5}{6} = \frac{9}{13}.$$

Much easier. And for those who would like to apply the method more generally, the formula is

$$\frac{a}{c} + \frac{b}{d} = \frac{a+b}{c+d}.$$

At this stage you may be suspicious. So, ok, we'll confess: this excellent method for addition is not really our invention. This method of "adding" fractions has been discovered and rediscovered by schoolchildren for centuries. Of course, there's at least one problem with the method: it usually gives the wrong answer.

So, we'll all still have to add fractions in the traditional manner, with those annoying common denominators. However, the weird addition above does turn out to have some very remarkable properties.

Here's an interesting experiment, perfect for a sunny spring day. (So, you may have to leave Victoria.) Buy some concentrated raspberry cordial and mix yourself a glass of cordial and water. Of course the more cordial you add, the redder the liquid. We'll now consider two different mixes, in equal-sized glasses:

First Glass: 4 parts cordial and 7 parts water.

Second Glass: 5 parts cordial and 6 parts water.

Now create a third mix by combining the contents of the two glasses:

Third Glass: 4 + 5 parts cordial and 7 + 6 parts of water.

So, the proportions of cordial and water in the third glass are exactly given by the result our weird fraction sum. Intriguing.

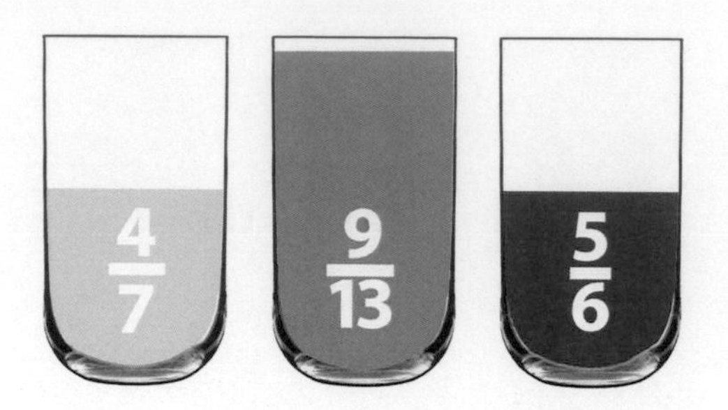

Now, since $5/6$ is greater than $4/7$, the second glass will be redder than the first. What about the third glass? Since we've just combined the contents of the first two glasses, the third glass will be in between, redder than the first but not as red as the second. That means we have a cordial-powered proof that

$$\frac{4}{7} < \frac{9}{13} < \frac{5}{6}.$$

Very neat.

The strange sum we've been considering is called a *mediant*, the name reflecting the in between property that we've just observed. Of course it's misleading to use a plus sign, and so our weird operation should be represented in some other way. Most commonly the symbol $\oplus$ is used. Then, for example,

$$\frac{4}{7} \oplus \frac{5}{6} = \frac{9}{13}.$$

However, there is something very strange about the mediant. To illustrate, first note that $5/6$ obviously equals $10/12$. However, if we calculate the mediant with $10/12$ in place of $5/6$ we find that

$$\frac{4}{7} \oplus \frac{10}{12} = \frac{14}{19}.$$

The two fractions $9/13$ and $14/19$ are definitely not equal. So, the mediant cannot be operating on the actual fractions, the numbers. Rather, the mediant is an operation on particular *representations* of fractions.

The mediant is definitely a peculiar creature, but it is still of genuine use. Our cordial mixing above is one illustration, but there are much more impressive applications.

Start with any positive whole number: we'll choose 6 to illustrate. Now write down, in order from smallest to largest, all the fractions with denominators at most 6; the fractions should be in lowest form, and we'll include $0/1$ and $1/1$ at the beginning and end. So, starting with 6, our list of fractions is

$$\frac{0}{1}, \frac{1}{6}, \frac{1}{5}, \frac{1}{4}, \frac{1}{3}, \frac{2}{5}, \frac{1}{2}, \frac{3}{5}, \frac{2}{3}, \frac{3}{4}, \frac{4}{5}, \frac{5}{6}, \frac{1}{1}.$$

Lists constructed in this way are called *Farey sequences.* These sequences have an amazing property: any fraction in a Farey sequence is the mediant of the fractions on either side. So for example, in our list above $3/5 = 1/2 \oplus 2/3$. That is very strange, and very, very cool.

It turns out that Farey sequences are much more than just weird fun. The *Riemann hypothesis* is perhaps the most famous and most important unsolved problem in mathematics (and is worth $1,000,000).[1] And, the Riemann hypothesis can be expressed as a question about Farey sequences.[2]

Amazing. And all that from a glass or two of cordial.

Puzzles to ponder

Can you find an example where the mediant of two "fractions", a/b and c/d, is equal to the sum of the two fractions? That is, your task is to find an example where

$$\frac{a}{c} + \frac{b}{d} = \frac{a+b}{c+d}.$$

Suppose now that a, b, c and d are all positive. Can you prove our cordial theorem? That is, assuming $a/b < c/d$, your job is to prove algebraically that

$$\frac{a}{c} < \frac{a+b}{c+d} < \frac{b}{d}.$$

[1]The Riemann Hypothesis is a "Millennium Prize Problem", with the Clay Mathematics Institute offering $1,000,000 for its solution. Burkard has a Mathologer video that explains some of the underlying mathematics.

[2]S. Kanemitsu and M. Yoshimoto, *Farey sequences and the Riemann hypothesis*, Acta Arithmetica, **75**, 363–378, 1996.

CHAPTER 2

Uncovering base motives

10.01021101222...

Once upon a time, many years ago, there was a review of the school mathematics curriculum. In those days it was believed that mathematicians could contribute some insight. The good mathematicians were pleased to assist, and the result was a brilliant curriculum. The teachers and students were delighted, and everybody learned mathematics happily ever after.

Well, not quite. Our tale is of the "New Math" movement, from the 1960s. The New Math did include a significant involvement of mathematicians, and the proposals were indeed mathematically sophisticated, but overall the results were farcical. It is perhaps the reason why mathematicians to this day reside in pedagogical purgatory.

A brilliant encapsulation of what went wrong is provided by Tom Lehrer's famous and funny song, *New Math.* Lehrer invites his audience to subtract 173 from 342. Having barely coped with that, Lehrer then declares that the subtraction should actually be done in base eight, chirpily singing that the 4 is in the "eights place", the 1 is in the "sixty-fours place", and so on. Lehrer's lesson is hilarious, full of New Math jargon, and all of it self-evidently pointless.

Arithmetic in anything but base ten has now disappeared entirely from the curriculum, and we bid good riddance to the nonsense lampooned by Lehrer. But why were these bases ever taught at all? Have we lost anything by their removal? Indeed, we have: the bases have been thrown out with the bathwater.

One purpose of using different bases is simply to have fun, to play with numbers. True, writing π in other bases may not appeal to everyone. We, however, enjoyed the exercise when trying to determine our ideal number plate.[1] But, apart from the games, an important message has been lost.

[1] In 2010, we wrote a column about wanting a customer car license plate, reading π in some suitable base.

Numbers are abstract. They are ideal, mental objects, difficult to discern and difficult to discuss. Because of this it is very easy, for example, to confuse the *numeral* "8", the symbol, with the *number* for which the symbol stands. Sadly, many current textbooks are riddled with such confusion.

Writing a number in different bases can be an attempt to distinguish the number from the symbols representing that number. Admittedly, the attempt may be so abstruse that the message is lost, and then we sing along with Tom Lehrer. However, given the current infestation of calculators, it is now common to view decimal representations as the be-all and end-all, to regard these decimals themselves as the numbers. They are *not* numbers, and they are usually not even insightful representations of numbers: this message is more important than ever.

That's all very general, so what about particular bases? The ancient inventor of our number system chose base ten simply because humans have ten fingers. In certain contexts, however, there are other natural bases. The clear example is base two, where all numbers are built up from 0 and 1. The on-or-off nature of base two arithmetic makes it perfect for the logic underlying computers, and for many related areas of mathematics. Indeed, given the technology fetishism of our curriculum masters, the absence of base two in the Australian curriculum is particularly puzzling.

Other bases have uses as well, and we'll end with a truly beautiful illustration. We have written before about irrational numbers,[2] and we remarked then that when numbers such as $\sqrt{2}$ are declared irrational it is usually with no hint of how we *know* them to be irrational. Well, now we will ponder that.

Recall that $\sqrt{2}$ being irrational means that it could not be written as a fraction, $\sqrt{2} = A/B$ with A and B whole numbers. If we square both sides of this equation, and multiply to get rid of the denominator, we then have the equation

$$A^2 = 2B^2\,.$$

So, what we are claiming is that this equation is impossible, that no positive whole numbers A and B will solve it. But how can we possibly rule out all of the infinitely many choices for A and B? Here comes the magic: we shall imagine that the numbers A and B are written in base three.

A number written in base three will have a ones place, a threes place, a nines place and so on. So, for example, we would normally write the number fifteen as 15, but in base three it would be written as 120: this amounts to $(1\times 9)+(2\times 3)+(0\times 1)$.

[2]See Chapters 33 and 53 of our first book of math columns, *A Dingo Ate My Math Book*, American Mathematical Society, 2017.

We can perform arithmetic in base three just as in base ten. For example, the base ten equation $2 \times 2 = 4$ would be written $2 \times 2 = 11$ in base three. And, here is the working for a harder one, the number 120 multiplied by itself:

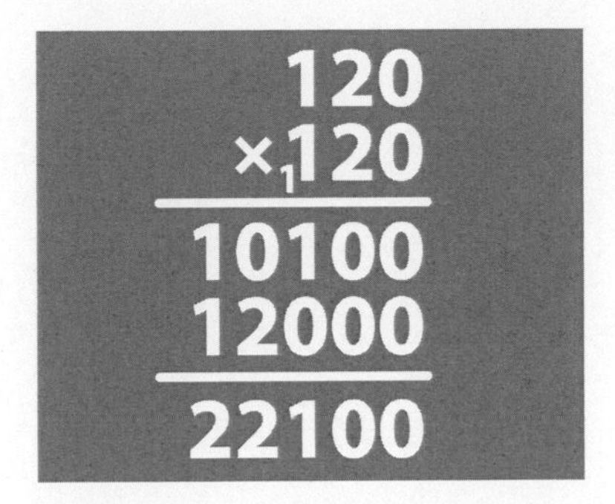

Of course you could convert 120 to base ten form (i.e. 15), and then calculate as usual, but the point is you don't need to: all the familiar methods of arithmetic apply just as well in base three, and in any base.

Now, for what follows, you only need to know one special fact about base three arithmetic:

Ignoring final zeroes, any squared number written in base three always ends in a 1.

For instance, reviewing our examples above, $1 \times 1 = 1$, and $2 \times 2 = 11$ and $120 \times 120 = 22100$. Why this is always so, why there is always a 1 at the end, may not be so obvious, although it is not that difficult to show. But hopefully the claim we're making is clear.

Now, with that fact in mind, look again at our A, B equation for $\sqrt{2}$. Written in base three, we now know that the A^2 will end in a 1 (ignoring zeroes). However, B^2 will also end in a 1, and that means $2B^2$ will end in a 2. But if A^2 and $2B^2$ end in different digits then they cannot possibly be equal. That is, the equation $A^2 = 2B^2$ is impossible to solve with positive whole numbers. And that means $\sqrt{2}$ cannot be written as a fraction. We have *proved* that $\sqrt{2}$ is irrational.

This amazingly simple proof is due to mathematician Robert Gauntt, who was a freshman at Purdue at the time.[3] Tom Lehrer is himself a mathematician, and we are sure that even the sceptical Lehrer would be captivated by this beautiful application of base arithmetic.

Puzzles to ponder

What is our alien up above doing?

Why does any squared number end in a 1 when written in base three?

Can you find another base, which proves that $\sqrt{3}$ is irrational?

[3] *The irrationality of* $\sqrt{2}$, American Mathematical Monthly, **63**, 247, 1956.

CHAPTER 3

Sneaky square dance

It is pleasing to know that there are parts of the world where everyone loves squares. What a paradise it must be, a country where people are forever marching in perfect square formation.

Well, maybe not. Geometric marching is probably not as much fun when its main purpose is to please a Glorious Leader. Still, those marching squares are impressive. And, given the unfortunate folk will be marching anyway, we have a great idea for a very mathematical flourish.

Our plan is to have two identical squares of marchers, each square performing the usual stunning steps. Then, the grand finale will consist of the squares being merged into one big (Red) square.

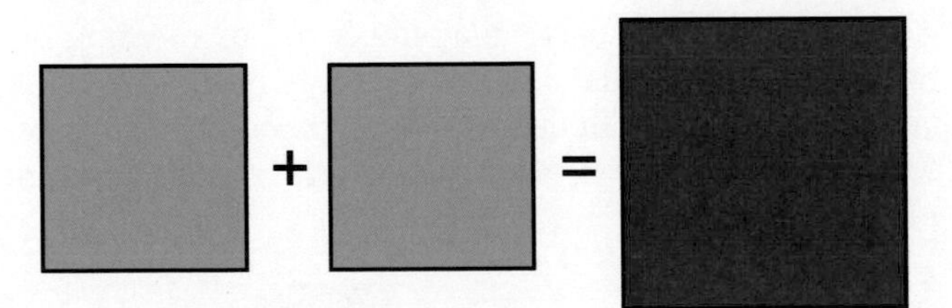

It'd definitely be a showstopper but first there are details to sort out. The interlacing of the squares will be tricky, requiring planning, practice and fancy footwork.

We also have to decide the size of squares to use, which would need to consist of a suitable number of marchers. For example, beginning with two tiny 2×2

squares wouldn't work: that would give us eight marchers in total, one short of the number required to rearrange into a 3×3 square. Similarly, beginning with two 3×3 squares of marchers would mean we have eighteen marchers, too many to form a 4×4 square and insufficient for a 5×5 square.

Hmmm. This will take some time to figure out, but we should be able to do it. We'll try 4×4, then 5×5 and so on, and eventually we should have in hand the smallest squares that work.

For now, let's leave that calculation and just assume we've determined the smallest squares that can be merged. Then, imagining we have sufficient marchers to occupy those squares, we can plan the marching steps.

Let's begin with an empty red quadrangle of just the right size to accommodate all our marchers. Then, a stylish approach would be to have the two identical squares of marchers enter the quadrangle from opposite sides.

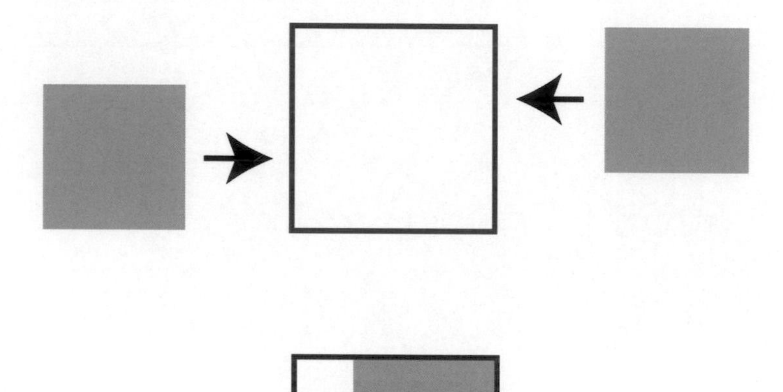

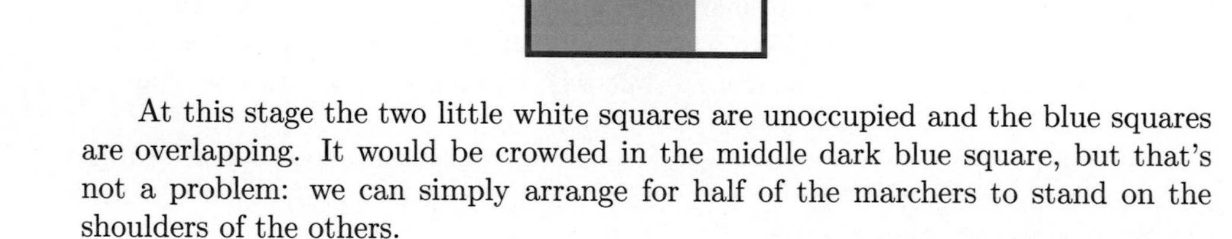

At this stage the two little white squares are unoccupied and the blue squares are overlapping. It would be crowded in the middle dark blue square, but that's not a problem: we can simply arrange for half of the marchers to stand on the shoulders of the others.

Finally, we'll have the marchers leap off their comrades' shoulders and into the two empty white squares, a spectacular finish to our merging of the two blue squares into the Red square. Ta da!

But wait a minute. If, as planned, we have precisely the correct number of leaping marchers to fill the two little white squares then this can be represented by the following picture.

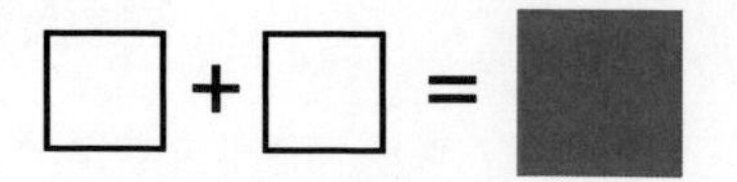

Uh oh. We assumed that we began with the very smallest squares that could be merged to make a larger square. Yet, somehow we created even smaller squares that would work. How can that be?

Simply, it cannot be. Sure, there is nothing wrong with our marching plan. However, for our *smallest* possible squares to result in even *smaller* squares is a plain logical impossibility. The unavoidable conclusion is that *no* squares can be merged in the way we had originally contemplated.

Well, bum. So much for our plans to impress our Glorious Leader with a great marching finale. However, perhaps he'll be impressed by some intriguing mathematics that emerges from our failed attempt.

What we have outlined above is a *proof by contradiction.* We began by *assuming* that certain squares were possible, and that assumption led to a *contradiction*, a logical impossibility. This contradiction *proves* that our original assumption was wrong, and that no such squares can exist.

We can now reconsider this conclusion in terms of numbers rather than squares. Consider again the two hypothetical blue squares. Say each square contains $B \times B$ marchers and that the larger red square into which they can supposedly merge has $R \times R$ marchers. We have proved this is impossible, and that means that there are no positive whole numbers B and R that solve the equation

$$2B^2 = R^2 .$$

Rearranging, it follows that there is no *fraction* R/B that solves the equation

$$\left(\frac{R}{B}\right)^2 = 2 .$$

So, there is no fraction whose square is 2, which is exactly the same as saying $\sqrt{2}$ is not a fraction. That is, our marching ponderings have *proved* that $\sqrt{2}$ is irrational.

Now that is pretty cool. It is one thing to punch $\sqrt{2}$ into a calculator and to stare at a few unilluminating decimals; it is another to *know* that $\sqrt{2}$ can never be written as a fraction.

We've written about $\sqrt{2}$ before, when discussing the clever principle behind the dimensions of A-sized paper.[1] On another occasion we used base arithmetic to give a different proof that $\sqrt{2}$ is irrational.[2]

There are a number of proofs that $\sqrt{2}$ is irrational, but the marching band proof above is possibly our favorite.[3] It has been popularized by the great John Conway, who attributed it to mathematician Stanley Tennenbaum.[4]

And will the Glorious Leader be impressed with Tennenbaum's proof? We don't know. But, if he isn't, and perhaps anyway, it may be time to start looking for a new Glorious Leader.

Puzzle to ponder

Can you come up with a similar marching proof that $\sqrt{3}$ is irrational?

---

[1] See Chapter 33 of *A Dingo Ate My Math Book.*

[2] See the previous Chapter.

[3] One of your Maths Masters leans towards the proof in the previous Chapter.

[4] J. Conway and J. Shipman, *Extreme Proofs I: The Irrationality of* $\sqrt{2}$, Mathematical Intelligencer, **35**, 2–7, 2013. We also generalized Tennenbaum's argument in *Marching in Squares*, College Mathematics Journal, **49**, 181–186, 2018.

CHAPTER 4

A very strange set of blocks

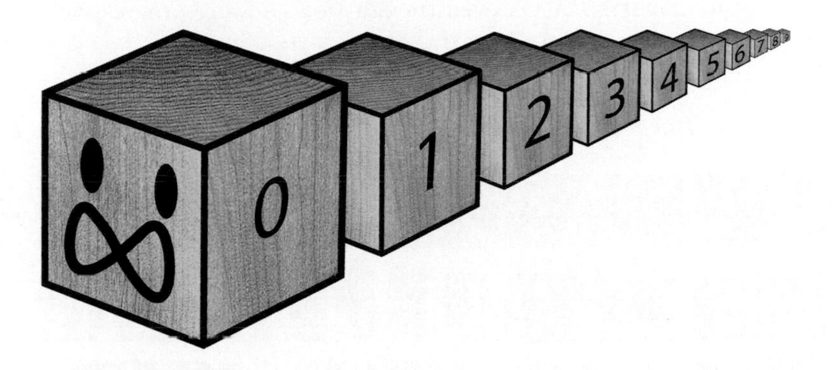

Your Maths Masters have been lecturing at universities for about three hundred years, and for the last two hundred years we have been engaged fulltime in popularizing mathematics. Something like that. Anyway, we have worked long and hard on many projects, but all our efforts have been guided by one fundamental goal: to convince as many people as possible that $0.999\cdots = 1$.

We look forward to the day when we can visit a school, ask the students what the infinite decimal $0.999\cdots$ is, and have them all shout back "One!" On that day we can happily retire. It is not likely to be soon.

Infinity is a very tricky concept, which has been bamboozling mathematicians for millennia. In fact it was only in the 19th century that infinite constructions, including $0.999\cdots$, were completely understood.

We have another infinity puzzler in store, so won't revisit $0.999\cdots = 1$ today.[1] We will pause, however, to note that an infinite decimal is in fact an infinite sum. Our friend $0.999\cdots$, for example, is the infinite sum

$$9/10 + 9/100 + 9/1000 + \cdots .$$

A similar and possibly more familiar infinite sum is

$$1/2 + 1/4 + 1/8 + \cdots .$$

This sum also totals exactly to 1.

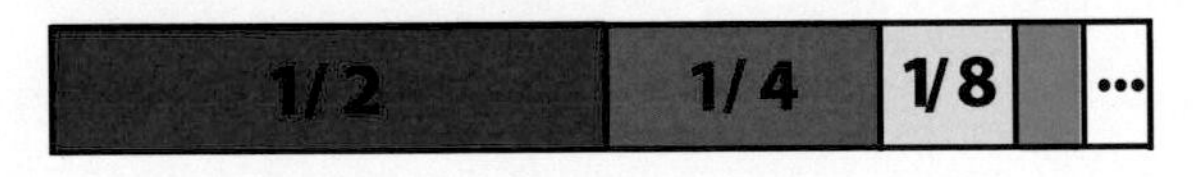

[1]See Chapter 48 of *A Dingo Ate Our Math Book*, and the puzzle for this Chapter.

A sum where each number is a fixed multiple of the previous number is known as a *geometric series*. That is true for both the sums above: at each stage we're multiplying by $1/10$ to create our old friend $0.999\cdots$, and by $1/2$ in the second sum.

The nice thing about a geometric series is that, just as for the above examples, the total can be worked out exactly. However, it will come as no surprise that infinite sums other than geometric series can be much trickier.

Consider the following sum, known as the *harmonic series*:

$$1 + 1/2 + 1/3 + 1/4 + \cdots .$$

What can we say about it? Well, even though we're adding progressively smaller numbers, the total is very large. Notice that $1/3 > 1/4$, and so

$$1/2 + (1/3 + 1/4) > 1/2 + (1/4 + 1/4) = 1/2 + 1/2 .$$

Not so large yet. But now have a look at this.

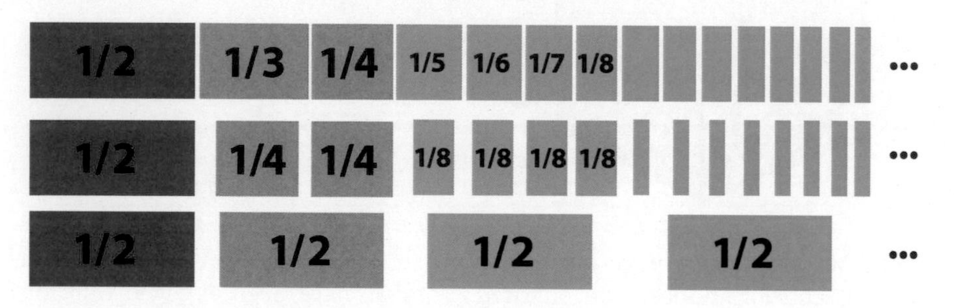

As pictured, the next four (blue) fractions also sum to at least $1/2$, as do the next eight (green) fractions, and so on, forever. So, since we're adding $1/2$ over and over forever, the only possibility is that the harmonic series totals to infinity. We told you it was large.

OK, now onto another sum, the *quadratic series*:

$$1/1 + 1/4 + 1/9 + 1/16 + \cdots .$$

This time the denominators are squares, and so the sum is smaller. But how small?

Imagine infinitely many squares, the first with side length 1, the second with side length $1/2$, then $1/3$, and so on. The areas of these squares are $1 \times 1, 1/2 \times 1/2, 1/3 \times 1/3 \cdots$. It follows that the sum of the areas of all the squares is exactly the quadratic series.

However, as the picture below illustrates, the initial 1×1 square is large enough to accommodate all the subsequent squares.

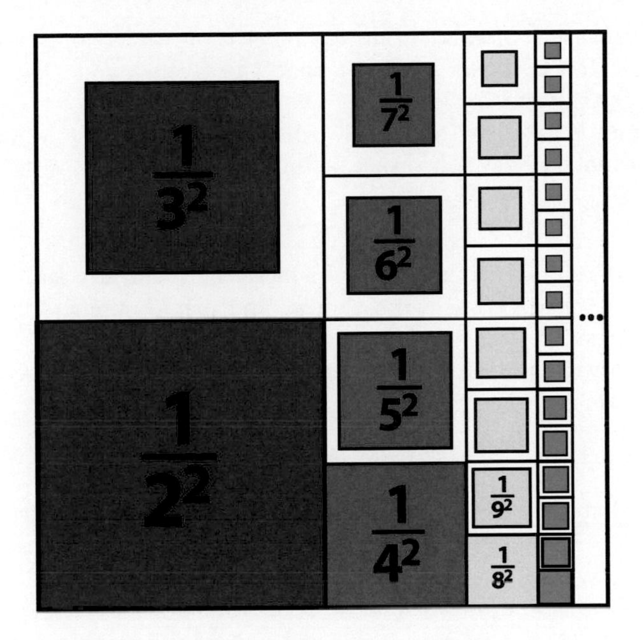

That means that the total sum of the areas is at most twice that of the first square. So, the quadratic series totals to less than 2: definitely finite.

We now want to consider a very puzzling toy, suitable for a budding baby maths master. The toy we have in mind is an infinite set of blocks: the first block has dimensions $1 \times 1 \times 1$, the second is $1 \times 1/2 \times 1/2$, the third $1 \times 1/3 \times 1/3$, and so on. Let's consider how we could make these blocks.

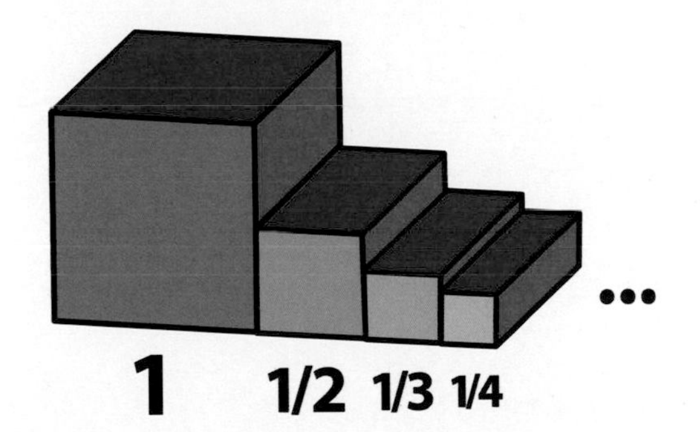

From the familiar "length times width times height", we can calculate the volume of each block. It is then straightforward to calculate that the sum of the volumes of all the blocks is exactly our quadratic series. That means the total volume of the blocks is less than two: it'd take a lot of work to make them, but we'd only require a finite amount of wood.

But, what if we wanted to paint the blocks? Just considering the red sides on top, the first has area 1, the second area 1/2, the third 1/3 and so on. That means the sum of the red areas is exactly the harmonic series, and so must infinite.

Hmm. These are truly strange blocks: it would take a finite amount of wood to make them, but the infinite surface area means it would require an infinite amount of paint to paint them.

But what if we just hollowed out the blocks? It would only require a finite amount of paint to fill them: wouldn't that effectively paint that same infinite area? It's all very strange!

One final matter: we have yet to say what the quadratic series actually totals to. Believe it or not, we have the exact, astonishing answer:

$$\frac{1}{1} + \frac{1}{4} + \frac{1}{9} + \cdots = \frac{\pi^2}{6}.$$

How on Earth did π get in there? In fact, the quadratic series is the starting point for many fascinating stories, on π, and prime numbers, and the famed *Riemann Hypothesis*.

We'd love to write on that sum one day.[2] But, first, there is some homework to be done: a little matter of $0.999\cdots = 1$.

Puzzle to ponder

Can you *prove* that $1/2 + 1/4 + 1/8 + \cdots = 1$? (Chapter 24 provides a hint.)

[2] Burkard has some *Mathologer* videos devoted to the sum and to the Riemann Hypothesis.

CHAPTER 5

Parabolic production line

One of your Maths Masters is blessed to have a very young maths mistress to assist him with his work. Regular readers of this column will not be surprised that little Eva's favorite word is "cat". She loves cats. Amusingly, though, it seems that everything is a cat. Eva greets all animals (including her parents) with a loud and confident cry of "Cat!"

Eva's omni-catting is oddly familiar. It is reminiscent of many high school math classes, where any curvy graph is as likely as not to be greeted with confident cries of "Parabola!"

The parabola-spotting is significantly less amusing, but of course it is not the students' fault. The textbooks overflow with parabolas and quadratic equations, including the most absurdly contrived applications: what to make of a shop displaying $x^2 + 2x - 48$ types of cheese?[1] Moreover, students are introduced to few other curves, which are seldom distinguished in a meaningful or memorable manner.

It is all sad and silly. And needless. There are genuine, beautiful applications of parabolas which are rarely if ever discussed. Yes, eventually a few lucky students briefly study projectile motion (and seemingly more briefly when the senior Australian Curriculum kicks in), but that's about it.

Physics students still see such applications, in the form of parabolic mirrors and lenses. These are welcome mainstays of science museums such as Scienceworks,[2] particularly as very impressive parabolic whispering walls.

Wouldn't it be refreshing for students to know how these displays work, beyond parroting "focal point" as if reciting jargon explains anything? Wouldn't it be worthwhile for students to know what a focal point is, and the mathematics to explain it? We live in (not overly much) hope.

---

[1]We're not making this up.

[2]Melbourne's not very good hands-on science museum, which we've had cause to mention a few times. See Chapter 40, and Chapters 63 and 64 of *A Dingo Ate My Math Book*.

One day we'll give focal points a go.[3] Today, however, we'll demonstrate a simpler and very cool feature of parabolas.

Last year, one of your Maths Masters visited the *Mathematikum* in Giessen, Germany. This is a seriously amazing mathematics museum, featuring many impressive exhibits, including a parabolic whispering wall. One exhibit was based upon a parabola drawn on a white wall:

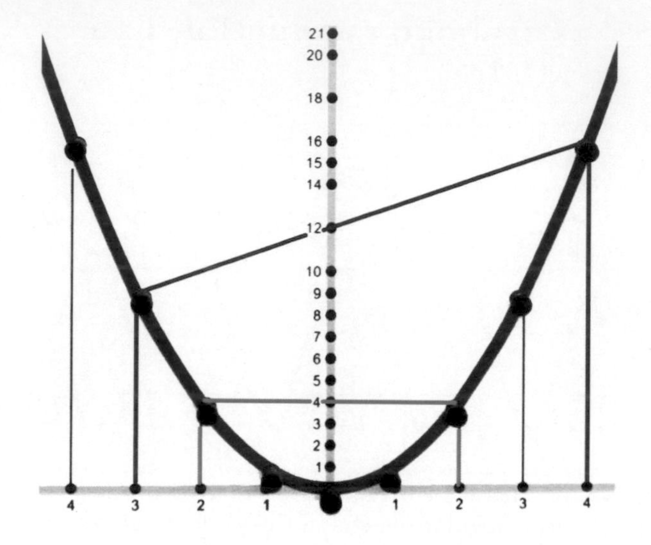

What appear as black dots on the parabola are cylinders protruding from the wall. The numbers on the coordinate axes indicate that the parabola is the archetypal $y = x^2$, with the cylinders placed at points with whole number coordinates, $(1, 1)$, $(2, 4)$ and so on.

The purpose of the exhibit is to demonstrate how to graphically multiply two numbers. To find 3×4, for example, the visitor is instructed to tie a rope between the cylinders at distances 3 and 4 to the left and right of the vertical axis. This rope is represented by the sloping red line.

Then, the spot where the rope crosses the vertical (y) axis indicates the product of the two numbers. So, in the above example we conclude that $3 \times 4 = 12$. Very nice.

It is a surprising and clever demonstration, but why does it work? It is not hard to understand when the two numbers are the same, amounting to squaring a single number: in this case the rope will be horizontal and at just the correct height. This is illustrated by the green line, indicating the product $2 \times 2 = 4$.

[3]See Chapter 64 of *A Dingo Ate My Math Book*.

If the numbers A and B to be multiplied are unequal there is more work to be done. A natural approach is to begin by obtaining the equation of the straight red line. However, it is simpler to just draw in horizontal and vertical lines, creating a big red triangle:

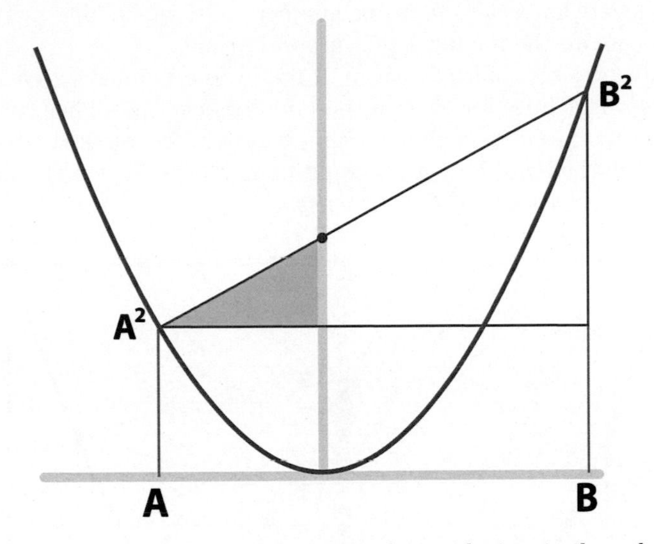

Then, we notice that the smaller triangle inside is *similar*, that is, it's the same shape. It is then straightforward to use this similarity to obtain an equation, indicating where the red line crosses the vertical axis. We'll leave the details to the triangle aficionados.

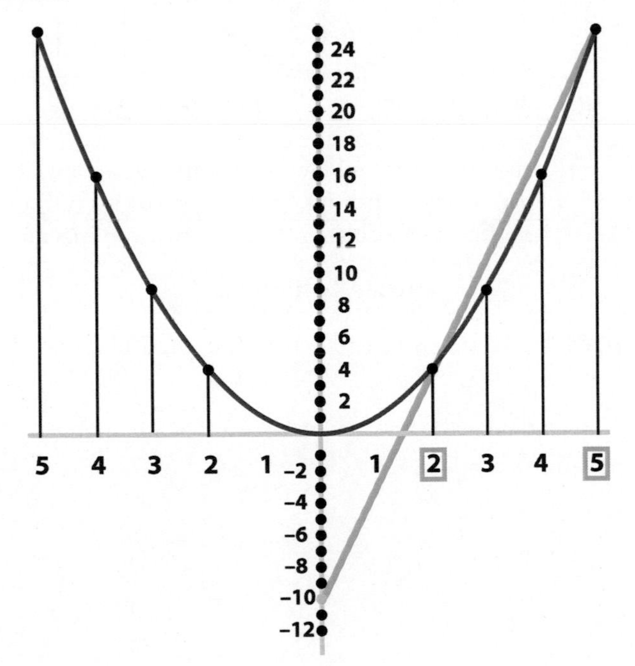

What about negative numbers? The Mathematikum exhibit only demonstrated the multiplication of positive numbers, but the same construction works just as well in general. Above, for example, we have calculated $2 \times (-5) = -10$.

This graphical multiplying is very pretty, and it also has a very pretty consequence: it can locate for us all the prime numbers. Or, which amounts to the same thing, it can locate all the numbers that are *not* prime.

Consider all possible multiplications of two whole numbers greater than one, and draw the straight lines for each of these multiplications. Then, every possible product will be indicated by a line running through the vertical axis. The only numbers left untouched will be 1 and the prime numbers. It is a striking, pictorial variation of the famous *sieve of Eratosthenes*.

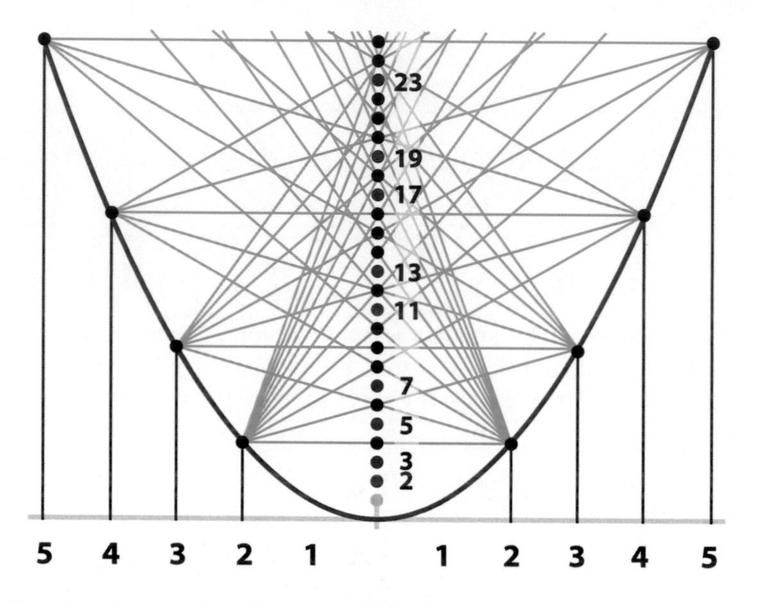

It's all very nice stuff and we could write plenty more, but that's probably enough parabolic fun for one day. After all, there's serious schoolwork to get done: somewhere out there, there is a cheese shop that desperately needs factorizing.

Puzzle to ponder

Use the pictured similar triangles to prove that the parabolic trick for multiplying really works.

CHAPTER 6

The magic of the imaginary

One of our favorite mathematical writers is John Stillwell, formerly at our home of Monash University, and perhaps our favorite among John's many excellent books is *Yearning for the Impossible.*[1] In this book John describes the manner in which mathematicians fantasize about the "impossible", and how they make these fantasies real. More bluntly put, John tells the story of mathematics as an extended history of cheating.

Probably the most notorious example of mathematical cheating is the introduction of *imaginary numbers.* The name hardly inspires confidence, and the alternative designation of "complex numbers" is not much better.

What are imaginary numbers? The common and unsatisfying answer is that they are, for example, square roots of negative numbers, $\sqrt{-1}$ and $\sqrt{-2}$ and so forth. The eminently reasonable complaint is that a number multiplied by itself cannot be negative, and so these weird roots simply don't make sense. The cheater's rejoinder is that he doesn't care whether or not such numbers exist; he's just going to cheat and pretend that they do.

The cheater's approach is to avoid asking what imaginary numbers *are*, or where we might find them, and instead focus upon what they *do*. That may seem intellectually dishonest but the approach is very common.

Let's step back and consider an everyday square root, an apparently harmless fellow such as $\sqrt{5}$. Even here it is very difficult to say what the number actually is. Sure, we can write $\sqrt{5} = 2.36067\cdots$, or whatever. However, writing out a few decimals followed by some dots is doing no more than giving the illusion of precision and understanding. It is extraordinarily difficult, however, to make sense of infinite decimals. Indeed, even apparently simple repeating decimals, such as $0.99999\cdots$, can be very tricky.[2]

[1] A K Peters, 2006.

[2] See Chapter 4.

Nonetheless, if we cannot say precisely what $\sqrt{5}$ is, we still know exactly how it works. The whole point of "root" is that it's the reverse process of "squaring". So, for example $\sqrt{4} = 2$ because $2 \times 2 = 4$. Similarly, for our troublesome $\sqrt{5}$ fellow, the one thing we can be sure of is

$$\sqrt{5} \times \sqrt{5} = 5\,.$$

Well, in for a penny, in for a pound. We may well not believe that $\sqrt{-2}$ exists. If however, we shut out our protesting brain and just pretend that it does, that there is something meaningful there, then the one equation we will accept is

$$\sqrt{-2} \times \sqrt{-2} = -2\,.$$

That's all good fun but it also appears to be tautological. Is there anything substantial to be gained? Amazingly, there is.

To illustrate, consider the following sequence of numbers:

$$1, 1, 2, 3, 5, 8, 13, 21, \cdots.$$

These are the megafamous *Fibonacci numbers.* They're not actually due to Fibonacci although, as we've discussed elsewhere, there are other good reasons to honor him.[3] So, we begin with 1 and 1, then $2 = 1 + 1$, $3 = 1 + 2$, and so on, each Fibonacci number being the sum of the two previous ones.

That's all well and good, but what if we want the 1000th Fibonacci number? Sure, we can churn the numbers out one by one: the 9th Fibonacci number is 34, the 10th is 55, and on and on. We'll eventually get there. But what we want is a formula that will simply provide us with the answer immediately. That's where the magic begins.

Fibonacci wrote down his famous sequence in 1202. The magic formula, now known as *Binet's formula*, came over 500 years later. Discovered by French mathematician Abraham de Moivre,[4] Binet's formula gives the 1000th Fibonacci number as

$$\frac{\left(\frac{1+\sqrt{5}}{2}\right)^{1000} - \left(\frac{1-\sqrt{5}}{2}\right)^{1000}}{\sqrt{5}}\,.$$

It is hard to exaggerate the amazingness of Binet's formula. Clearly the Fibonacci numbers will all be whole numbers. Nonetheless, Binet's formula enters the world of irrational numbers and then exits again. All we need use, and all we can use, is that $\sqrt{5} \times \sqrt{5} = 5$, and all the irrational roots magically cancel out.

(As an interesting side point, notice the golden ratio $(1 + \sqrt{5})/2$ appearing in Binet's formula. It is pleasing to see the golden ratio taking time away from the tawdry business of selling cars to perform an honest day's mathematical work.[5])

We can perform exactly the same magic with imaginary numbers. To illustrate, let's consider another sequence:

$$1, -1, -5, -7, 1, 23, 43, \cdots.$$

[3]We wrote a column based upon Keith Devlin's excellent book, *The Man of Numbers.* Devlin focusses on the importance of Fibonacci's work to the rebirth of mathematics in Europe.

[4]Binet's formula is one of the *many* examples of a formula or theorem being named after someone other than the discoverer. See Chapter 58.

[5]See Chapter 59 of *A Dingo Ate My Math Book.* In this chapter we hammered the use of the golden ratio as a sales gimmick.

Known for historical reasons as the *Marty numbers*,[6] they are deservedly much less famous than Fibonacci's sequence. They are constructed, however, in a very similar manner.

For the Marty numbers, we begin with 1 and -1. Then, any subsequent Marty number is twice the previous number minus three times the one before that. So, the third Marty number is $(2 \times -1) - (3 \times 1) = -5$. The 8th Marty number would be $(2 \times 43) - (3 \times 23) = 17$, and so on.

And now the question: what if we want the 1000th Marty number? Here it is:

$$\frac{(1+\sqrt{-2})^{1000} + (1-\sqrt{-2})^{1000}}{2}.$$

And, see where we've ended up. Whereas the Fibonacci numbers required us to consider the irrational, the Marty numbers have led us all the way into the imaginary world. Once again, all the roots magically cancel out, giving an ordinary everyday whole number, and to accomplish this all we need is to apply the equation $\sqrt{-2} \times \sqrt{-2} = -2$.

Not that the Marty numbers are particularly interesting; it is only your Maths Masters and a few of their mates who have ever bothered with them. But the Marty numbers enable a simple and historically faithful demonstration of how to cheat with imaginary numbers, how to extract a real, workable answer from these weird, semi-real creatures.

The practice of employing imaginary numbers to solve problems about everyday numbers has a proud and puzzling history. It began in the 1500s, when Scipione del Ferro and other Italian mathematicians discovered a formula for the solutions to cubic equations, the higher degree counterpart to quadratic equations. Their formula always worked, but sometimes in a perplexing manner: even if the solutions of a polynomial were everyday numbers, the formula might express these solutions in terms of imaginary numbers; just as happened with the Marty numbers. Del Ferro and his colleagues had no idea what these imaginary numbers were or why they were necessary. They just knew that they worked.

Cheating with imaginary numbers continued for centuries. The finest mathematicians, including the great Leonhard Euler, became masters at manipulating imaginary numbers without ever knowing what these numbers were or whether they existed. There is a sense in which *Euler's formula*, the gem of imaginary numbers, was never properly understood by Euler.[7]

Finally, in 1799 the Norwegian mathematician Caspar Wessell explained it all. Wessell provided a convincing explanation of what imaginary numbers actually are, making them as tangible, as "real", as everyday numbers. (They are arguably even more real: it was another 50 years before "harmless" irrationals such as $\sqrt{5}$ were satisfactorily explained.)

So, did the cheating end? Yes and no. By the end of 19th century mathematicians were much clearer on the rules of the game. It was no longer permitted to blithely concoct new numbers without a solid sense of what these new numbers were. Nonetheless, new numbers and whole new mathematical worlds were, and are, still being concocted.

[6]Patent pending.

[7]Euler's formula states that $e^{\pi i} + 1 = 0$, where i stands for $\sqrt{-1}$ and e is *not* "Euler's number": see the next Chapter.

Mathematicians continue to just make things up, just as they always have. There is a famous quote by mathematician Leopold Kronecker:

God made the integers; all else is the work of Man.

Indeed, Kronecker doesn't go far enough. Nobel prize winning physicist Percy William Bridgman responded to Kronecker's quotation:

Nature does not count nor do integers occur in nature. Man made them all, integers and all the rest.

Bridgman was correct. All mathematics is cheating. It's all a fiction.

Puzzle to ponder

Write $\phi = \frac{1+\sqrt{5}}{2}$ for the *golden ratio*, and write $\mu = 1 + \sqrt{-2}$ for the (just now coined) *Marty ratio*. Prove that

$$\phi^2 = \phi + 1$$

and

$$\mu^2 = 2\mu - 3\,.$$

(These equations imply that the sequences $1, \phi, \phi^2, \phi^3, \cdots$ and $1, \mu, \mu^2, \mu^3, \cdots$ follow, respectively, the same pattern as the Fibonacci sequence and the Marty sequence. That, in turn, is the key to proving Binet's formula and (the just now coined) Marty's formula.)

CHAPTER 7

There's no *e* in Euler

Recently, while admiring Melbourne's gorgeous display of overhead wires,[1] we had cause to mention the number e. We also remarked that common references to e as "Euler's number" were inaccurate. (And, no, that's *not* the great Swiss mathematician pictured above). Some of our readers queried that claim. We'll now reply, taking the opportunity to tell a small part of the story of e.

Though extremely important, e is not the most inviting of numbers. Unlike π, the number cannot be illustrated or motivated or explained by very simple geometry: we cannot simply point to a circle or similar and exclaim "Look, there's e!"

This difficulty of e is exemplified in the Victorian curriculum. The curriculum exhibits no concern for what e is, or why it is what it is. It also appears that the forthcoming Australian curriculum is inclined to do little more. Although e requires some effort, however, it is not nearly as difficult a number as is suggested by this dereliction of duty.

The historical origins of e are clouded by the mists of time, but it seems likely that the number first arose as the result of financial considerations. It is still the easiest way to get a grasp of the number.

To begin, imagine we come across a very generous bank, Bank Simple, offering 100% annual interest. (Yes, this is a fantasy.) So, if we invested \$1 then after a year our dollar would have doubled to \$2.

[1] See Chapter 34.

That is a very good offer, but then we find a second bank, Bank Compound, with a different scheme: they will give us 50% interest every six months. Would we prefer to invest with Bank Compound? Definitely.

After six months at Bank Compound, we will have 50% on top of our original \$1, amounting to \$1.50; in effect, we've multiplied by 1.5. Then, after the next six months, we will have earned an additional 50% of that \$1.50: we again multiply by 1.5, to arrive at the year's total of \$2.25.

This is illustrating the familiar and important notion of *compound interest*. The point is that calculating smaller interest at correspondingly smaller time intervals means that we are obtaining interest on our interest, resulting in a greater overall return on our investment. What does this have to do with e? We're getting there.

Imagine we've found a third bank, Bank Super Compound, which returns 25% interest every three months. We would then obtain a 25% increase in our investment, compounded four times over the year. So, starting again with our faithful \$1, at the end of the year we would have $(1 + 1/4)^4 = \$2.44$.

We can keep going. We locate Bank Super Duper Compound, which calculates the interest every month, contributing an extra 1/12 on top of our investment on each occasion. The result is, at the end of the year our \$1 would have grown to $(1 + 1/12)^{12} = \$2.61$.

Finally, we come across Bank Infinity, which goes the whole hog. This last bank divides the year into a zillion nanoseconds and then calculates the appropriately tiny amount of interest at each nanosecond. The result is, at the end of the year our \$1 will have returned $\$(1 + 1/\text{zillion})^{\text{zillion}}$. (To be precise, Bank Infinity calculates the *limit* of this quantity as the number of time intervals goes beyond a zillion and tends to infinity).

What is the amount returned by Bank infinity, what is this final number? It is the number we now denote by e. It is the result of compounding interest to the theoretical limit, what is known as *continuously compounded interest.* For those who love decimals, or calculating their interest *really* precisely, the expansion of this special number begins

$$2.718281828459045235360287471352662497757247093699 9 \cdots .$$

Do those final 9s indicate that the number is actually a terminating decimal? No: we've simply been cheeky in choosing where to stop. As is π, our new special number is irrational.

But what does any of this have to do with Leonhard Euler? Nothing.

The earliest known appearance of the number is an indirect reference in a 1618 work, probably by the English mathematician William Oughtred: it is Oughtred's portrait that we have included above. Our new number was then used throughout the 17th Century. Around 1690, the great German mathematician Gottfried Leibniz explicitly denoted this number by the letter b. That was seventeen years before Euler's birth, in 1707.

There is a second part to the story, however, which does involve Euler. Though we have indicated how Oughtred's number naturally arises as a financial concept, this only begins to explain the central, critical role that the number plays in calculus. This was demonstrated by Sir Isaac Newton and the other great 17th Century mathematicians.

This work on calculus was then carried to brilliant extremes by, among others, Leonhard Euler. And, along the way, Euler chose to denote Oughtred's number by the letter e, a labeling that has endured. There is absolutely no evidence, however, that Euler chose the letter e to refer to himself, or for any particular reason other than that the letter was relatively unencumbered by uses in other contexts.

Now, exactly how does e claim such a central role in calculus? And how, if at all, does e capture the notion of exponential growth? This is yet another issue ducked by our curricula. The question is simply handballed to the universities, which then typically drop the ball. Someday we'll get to that, too.[2]

Puzzle to ponder

Show that any compounding bank like the ones above will return at least \$2 by the end of the first year. That is, show that the quantity $(1 + 1/N)^N$ is always at least 2. (Since Bank Infinity is the "limit" of what these banks offer, this also proves $e \geqslant 2$. See the puzzle solution for a quick discussion on how to similarly go about proving $e \leqslant 3$.)

[2]Burkard's *Mathologer* YouTube channel has a video devoted to everything about e.

CHAPTER 8

What's the best way to lace your shoes?

Recently, we (possibly) figured out a shortest tour of Victoria.[1] Today we'll take you on some different tours, much closer to home.

When you lace and tie one of your shoes, the shoelace takes a tour of the eyelets. Two of the most popular such tours are the crisscross and zigzag lacings. However, there are many other possible lacings.

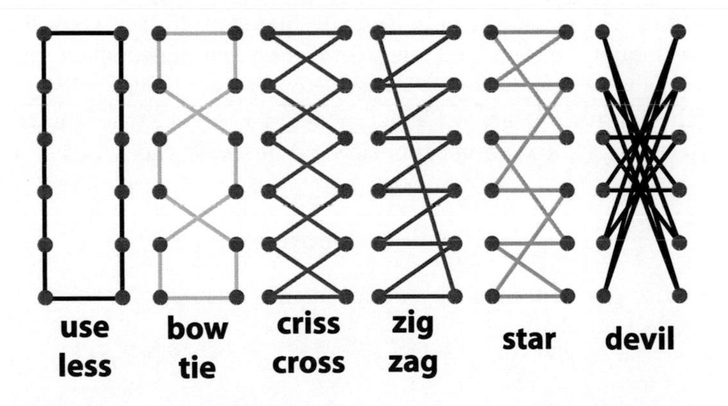

Which among these tours give the best lacing? To answer this we must first decide what we mean by "best". The simplest notion to capture mathematically is "best = shortest". So we'll begin by considering the lengths of lacings.

It is quite obvious that the lacing on the left is the shortest, while also being close to useless. We can preclude such short but silly lacings by requiring that

[1]See Chapter 44 of *A Dingo Ate My Math Book* where we used the idea of a trip around Victoria to discuss the traveling salesman problem.

each eyelet actually contributes to pulling the two sides of the shoe together. This amounts to having one or both segments that end in the eyelet connecting to an eyelet on the other side of the shoe. This is the case for all but the first lacing pictured above.

It turns out that, no matter the dimensions of the shoe and no matter the number of eyelets, the shortest *useful lacing* will always be the bowtie. Moreover, the lacings shown above are always in that order, from shortest to longest.

These are actually very surprising statements. For example, for a shoe with six pairs of eyelets there are 3,758,400 different useful lacings to compare. For God's Shoes, with 100 pairs of eyelets, the number of different useful lacings has grown astronomically, to:

1860904268074008196621915961820296614767464863477661360942284499556517
44439776524407967069878507881476622415958837227535961285224560640376326
00625910253777564367173849849651318613992974488058381689718183963494277
31395000391555199195008915051014033339502366067550230464677872323986936
3765334033228404327710700000000000000000000000000000000000000000000000

In most familiar lacings there are no vertical segments at all, with every segment connecting opposite sides of the shoe, which makes these *very useful lacings*. Among all such lacings, the crisscross is always the shortest possible, and the devil lacing shown on the right is the longest possible.

There is an alternative notion of "best lacing", which is perhaps more natural. We can view a lacing as a pulley system, pulling the two sides of a shoe together. We can then compare the strength of different lacings. Here there is no clear winner: depending upon the dimensions of the shoe, either the crisscross or the zigzag is the strongest. For dimensions close to those of real shoes – although who cares about those? – these two lacings are about equally strong.

So it turns out that the very popular crisscross scores well in terms of both length and strength. In addition, it is easy to remember, symmetric and pretty. It is fair to conclude that crisscross is indeed the best way to lace your shoes.

Finally, we should mention that the world's two leading shoelace experts reside in Melbourne: your writer here, Maths Master Burkard, who did the math for an article in the journal *Nature* in 2002;[2] and Ian Fieggen, who knows absolutely everything there is to know about shoelaces. You *must* check out Ian's amazing website.[3]

Puzzle to ponder

Count the number of useful and very useful lacings with 2×3 eyelets.

[2] *Nature*, **420**, 476. And, the real shoe fanatics can check Burkard's *The Shoelace Book*, American Mathematical Society, 2006.

[3] At the time of writing, Ian's website is still active.

CHAPTER 9

Ringing the changes

Do you like your math in exotic locations? Then why not join the band of bell ringers at Melbourne's St. Paul's Cathedral for the ringing of the changes. What does bell ringing have to do with math? A lot! We'll explain.

St. Paul's – the church opposite Federation Square – has twelve main bells, tuned in the key of C# major. The smallest bell, no. 1, rings the highest note, with the largest bell, no. 12, ringing the lowest. Changes can be rung using any number of bells. So, let's simplify things, and just use bells 1, 2, 3, and 4.

A *change* is what mathematicians call a *permutation*, the ringing of each of the four bells exactly once. For example, 3214 refers to the change of ringing bell 3, then bell 2, then bell 1, and finally bell 4. Then, *ringing the changes* means to ring a sequence of changes, whilst obeying three mathematical rules:

- First, the sequence starts and ends with the change 1234;
- Second, except for the start and end, no change is repeated;
- Third. from one change to the next, any bell can move by at most one position in its order of ringing. For example, this third rule says that the change 3214 cannot be followed by 2134, since bell 3 would have shifted by two spots.

One possible sequence of changes, known as *Plain Bob*, goes like this:

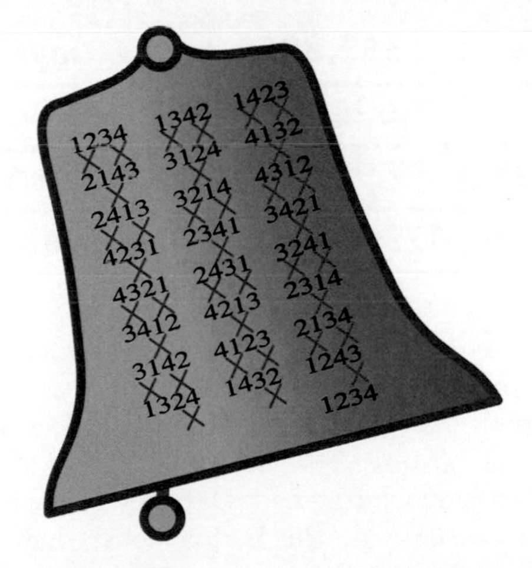

Here we first move down the first column, then the second and finally the third. Have a close look and you can see the third rule in action – from one change to the next a bell will either stay in the same position or swap its place with a neighboring bell. In the diagram, these swaps are indicated by crosses between the two changes.

Here are a few facts to set things into perspective. To ring one change takes between 1.5 to 2.5 seconds, the time it takes a bell to complete a natural swing.

When bell ringers go wild, they will ring sequences of changes consisting of more than 5000 changes, which translates into several hours of amusement for the neighbors. When ringing the changes, tradition dictates that bell ringers are not allowed any memory aids such as sheet music, nor can they be relieved by another bell ringer (for example, to relieve themselves). So, this means that a bell ringer has to effectively recite a sequence of several thousand numbers, one every two seconds, and to translate this sequence into perfect bell ringing. It takes a bell ringer several months to master ringing a bell by themselves, and years before they can dream of performing this kind of marathon bell ringing as a member of a team.

When ringing bells, one of the grand aims is to ring a sequence that includes every possible change. In the case of four bells such a sequence must be $(1 \times 2 \times 3 \times 4) + 1 = 25$ changes long, and Plain Bob is such an example. The general formula for n bells tells us that $(1 \times 2 \times 3 \times \cdots \times n) + 1$ changes are required.

Mathematicians have only recently proved that, no matter how many bells we want to ring, it is always possible to compose a complete ringing sequence.[1] Here is a table that shows you the numbers of changes we're talking about, and how long it would take if you rang at a rate of two seconds per change.

3	Singles	7	13 seconds
4	Minimus	25	49 seconds
5	Doubles	121	4 minutes
6	Minor	721	24 minutes
7	Triples	5,041	2 hours 48 minutes
8	Major	40,321	22 hours 24 minutes
9	Caters	362,881	8 days 10 hours
10	Royal	3,628,801	84 days
11	Cinques	39,916,801	2 years 194 days
12	Maximus	479,001,601	30 years 138 days

To ring a complete sequence on eight bells is the most that seems humanly possible. In recorded history, such a sequence has been rung on church tower bells only once. This took place at the Loughborough Bell Foundry in the U.K., beginning at 6.52 a.m. on 27 July 1963 and ending at 12.50 a.m. 28th July after 17 hours 58 minutes of continuous ringing. (Did they really not go to the toilet for 18 hours?!) Of course, to actually do this is ridiculously hard, both physically and mentally. So how do they do it? One mental trick is to make up sequences that are as easy as possible to remember. If you have a close look at Plain Bob, you can see that each column is generated from the change at the top by a simple knitting pattern of swaps. Then, at the end of the first and second columns, you swap the

[1] Arthur T. White, *Ringing the changes II*, Ars Combin, **20 A**, 65–75, 1985.

last two bells and in this way link the three columns together. Based on this simple algorithm, it is very easy to reconstruct the whole sequence from scratch. Try it!

Permutations and collections of permutations play very important roles in many branches of mathematics. In particular, *group theory*, the branch of mathematics concerned with symmetries, is full of permutations. The earliest examples of group theory structures and techniques in action are the highly structured bell-ringing sequences that were developed in the early 17th century. For example, the first column of our Plain Bob sequence captures the eight symmetries of the square. The subdivision of Plain Bob into three columns is something that mathematicians also get excited about.

But why on Earth did anybody dream up this convoluted mathematical way of ringing the bells, instead of just playing tunes? And, how did they get away with doing so for centuries? After all, lots of people do complain about the noise, and together with the bell tower itself the bells do form a musical instrument that you could in theory play tunes on by striking the bells with hammers.

The problem is that if you want to ring the bells by swinging them, which sounds a lot more impressive, and carries a lot further, ringing tunes is not an option. Why? The reason is that we are talking about very large bells, up to 1.5 tons in the case of St Paul's. Once set in motion it is very hard for a bell ringer to vary the interval at which such a monster bell will ring. This is the mechanical constraint that explains the third rule of bell ringing, and motivated bell ringers to invent mathematically perfect bell ringing.

Originally, change ringing was a competitive sport, with bands of ringers of footy player physique and mindset, competing against each other on a regular basis. Often at odds with the church itself, the *exercise*, as it is traditionally called, had as much do with shouting each other drinks according to very intricate penalty system for mess-ups, as it did with ringing the bells.[2] Bell ringing has come a long way and mellowed a bit since its bloodsport origins.[3] However, if you are interested in math, beer and serious mind games, you must visit one of six bell towers in Melbourne where bell ringing is still practised.

Puzzle to ponder

List all complete ringing sequences for three bells.

[2]In Australia, buying a friend an alcoholic drink is referred to as shouting.

[3]For more information, look up the Australian and New Zealand Association of Bellringers.

Part 2

The Shape of Things to Come

CHAPTER 10

Triangle surfer dude

Your Maths Masters are always on the lookout for engaging mathematics and yet again Federation Square suggests something worth pondering.[1] Take a look at the triangular window in the middle of the above photo; it is close to being equilateral and is neatly subdivided by the bars extending at right angles from the edges. There appears to be something there, but what?

To explain we'll revisit a classic textbook. Harold Jacobs' *Geometry* first appeared in 1974 and puts current texts to shame; it is fun, elegantly written and it contains a wealth of beautiful mathematics.[2] *Geometry* also contains many excellent problems and the very first problem concerns a marooned surfer.

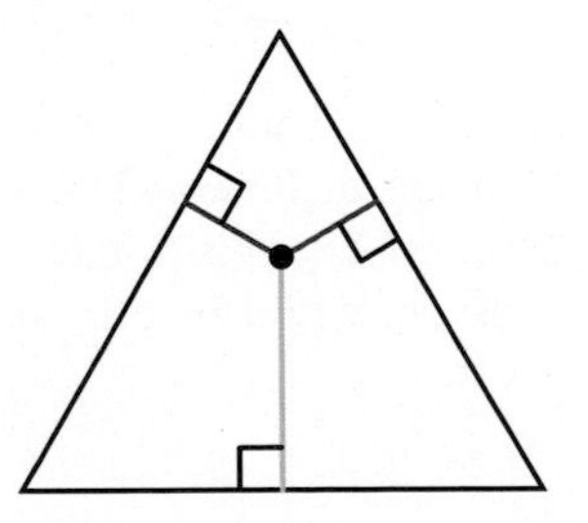

[1]Melbourne's cementy but very mathematical town square. See Chapters 18 and 19 of *A Dingo Ate My Math Book.*

[2]A third edition, published by W. H. Freeman, appeared in 2003. Our preference is for the first edition, also from W. H. Freeman, but any edition is infinitely better than any recent textbook.

Jacobs imagines a surfer stranded on an island in the shape of an equilateral triangle. (Happens all the time.) Our surfer dude enjoys all of the island's beaches and plans to spend an equal amount of time at each. That suggests that his hut, if he can be bothered to build it, would be most conveniently located so that the *sum* of the distances to the three beaches is as small as possible. Where would that be?

The surprising answer is that it doesn't matter: from any point inside an equilateral triangle the sum of the distances to the three sides is simply the height of the triangle. This lovely result is known as *Viviani's theorem*, named after the 17th Century Italian mathematician Vincezo Viviani.[3]

There are many proofs of Viviani's Theorem, and Jacobs' exposition is excellent. Perhaps the most beautiful, however, is a "proof without words" due to Japanese mathematician Ken-ichiroh Kawasaki. We've reproduced Kawasaki's proof below: just follow the simple mathematical steps from diagram to diagram.

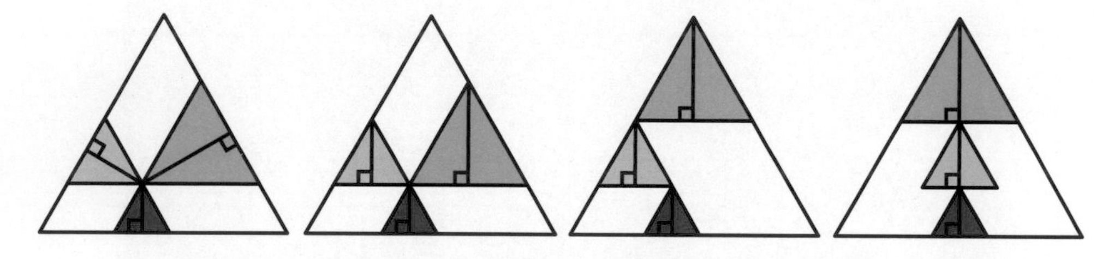

Unfortunately, our Federation Square window is not precisely equilateral. It consists of two of Fed Square's very special right-angled triangles, side by side,[4] making the top angle about 53°.

What if our surfer's island was the shape of the window? That's a more complicated situation but an elaboration of Kawaski's argument provides the answer: to minimize the sum of the distances the surfer can build his hut anywhere along the base of the triangle. (The answer would be different again if the top angle were greater than 60°.)

[3]Viviani is best known as Galileo's assistant and biographer.

[4]The sides of Federation Square's right-angled triangles have the proportions $1 : 2 : \sqrt{5}$. For why these proportions are special, see Chapter 18 of *A Dingo Ate My Math Book*.

Of course the above is all very contrived, since a surfer is not particularly likely to be marooned on a triangular island. He's obviously just as likely to be stuck on a quadrilateral or pentagonal or hexagonal island.

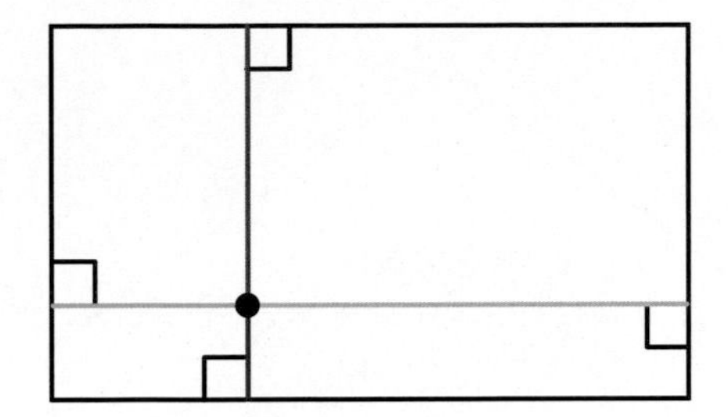

Luckily there are many different islands that possess the build-where-you-like property, and simplest of all are the fabled rectangular islands. On a rectangular island it is completely clear that the surfer can happily build his hut anywhere. What is less clear is that the same is pretty much true for any equiangular polygon.

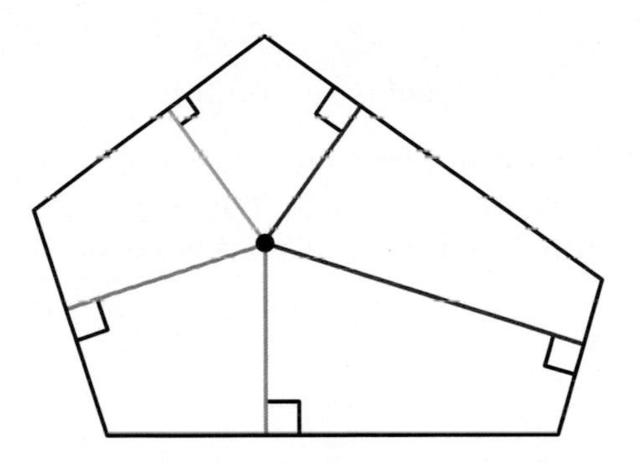

Consider the equiangular pentagon above, with each internal angle 108°. For many points on the island – the points from which every beach can be reached directly at right angles – the sum of the distances will be the same. A potential problem, which can arise if the surfer builds his hut too close to a beach, is that one of the natural perpendicular directions may instead take the surfer out into the ocean.

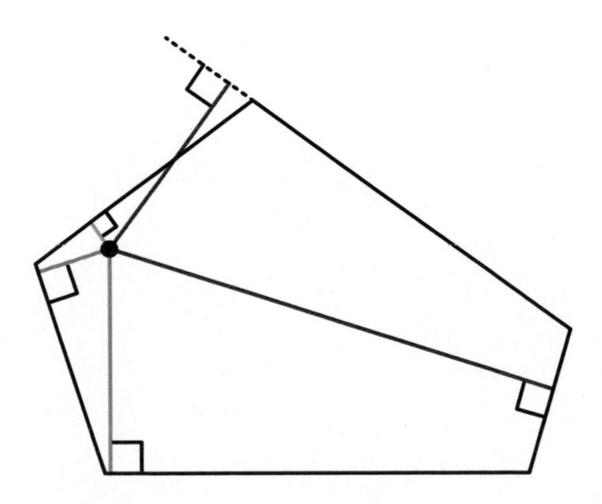

That is, however, the only thing that can go wrong with an equiangular polygon. A skinny island may have just one best spot on which to build a hut, but a more roundish equiangular island will offer a whole region of best spots from which to choose.

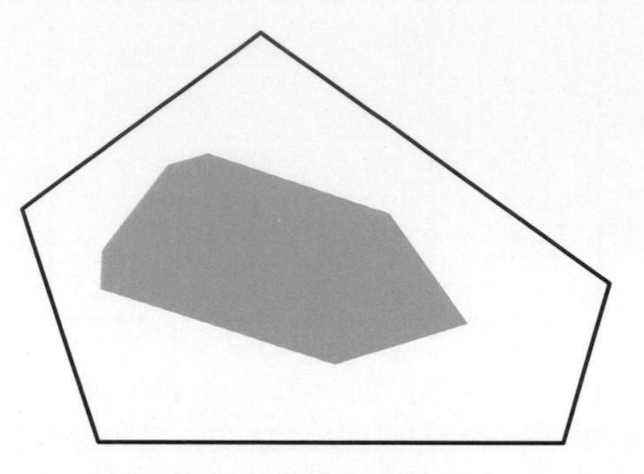

So, even a mathematically challenged surfer dude has a pretty good chance of getting it just right on an equiangular island. Life is good in Mathematical Surfer Land.

Puzzles to ponder

Show that on a *regular* pentagon island (all beaches are of equal length), the sum of the distances to the beaches is the same from any nice point. Using this fact, can you show that a general equiangular pentagon has the same property?

CHAPTER 11

We have it pegged

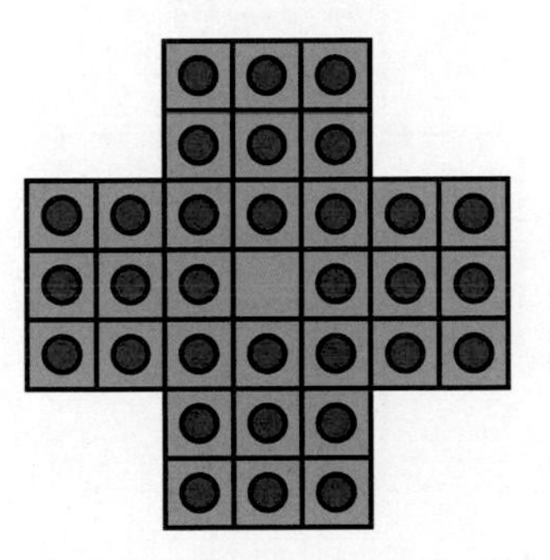

The other day a very excited little maths mistress returned home from school carrying a "brand new game". A quick inspection, however, revealed it to be a very old game: *Peg Solitaire.* Still, it turns out that there are interesting new things to say about this very old game.

For those unfamiliar with Peg Solitaire here is a quick reminder of the rules. The game begins with every space except for the center occupied by a marble, or peg, as pictured above. A move in the game consists of one marble jumping horizontally or vertically over another marble into an adjacent space, and then removing the jumped marble.

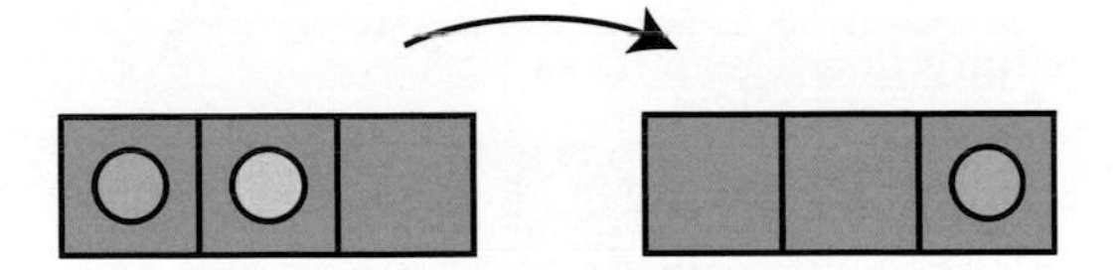

The goal of the game is to finish with just one marble, in the middle of the board.

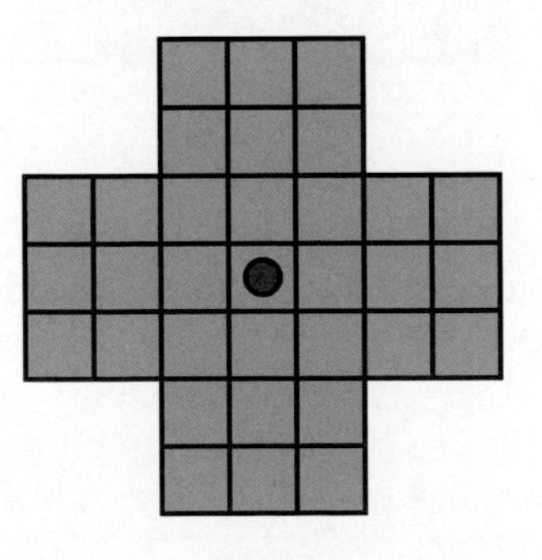

Before reading further the reader may wish to try the game for a few minutes. Or hours. (Online versions of the game are very easy to find.)

Peg Solitaire is not easy. Typically we'll be caught with two or more marbles and no possible moves. And, even if we succeed in removing all but one marble, chances are this last marble will not be located in the centre.

We won't tell you how to win the game; that'd spoil the fun, and anyone who really feels the need to cheat can easily find recipes online. Instead we'd like to point out and explain a surprising observation: if the game ends with just one marble remaining then that marble must be in one of the five spaces pictured below.

Why would that be so, and how could we possibly hope to prove it except by playing through every possible game? With the right approach it all turns out to be quite easy and very pretty.

To investigate the game we'll employ the gadget pictured below. It consists of two light bulbs, labelled Left and Right, each controlled by an on-off button.

We also label the solitaire board as follows:

<table>
<tr><td></td><td></td><td>L</td><td>LR</td><td>R</td><td></td><td></td></tr>
<tr><td></td><td></td><td>LR</td><td>R</td><td>L</td><td></td><td></td></tr>
<tr><td>L</td><td>LR</td><td>R</td><td>L</td><td>LR</td><td>R</td><td>L</td></tr>
<tr><td>LR</td><td>R</td><td>L</td><td>LR</td><td>R</td><td>L</td><td>LR</td></tr>
<tr><td>R</td><td>L</td><td>LR</td><td>R</td><td>L</td><td>LR</td><td>R</td></tr>
<tr><td></td><td></td><td>R</td><td>L</td><td>LR</td><td></td><td></td></tr>
<tr><td></td><td></td><td>L</td><td>LR</td><td>R</td><td></td><td></td></tr>
</table>

Now, whenever a marble is either removed from or placed on an L-space, we push the L button. Similarly, the R-spaces are our triggers to hit the R button, and the LR-spaces command us to hit both buttons. So, what can happen?

Let's begin with the solitaire board entirely empty and with both lights *off*. Now we add marbles one by one, switching the lights on and off as we go, until we're at the starting arrangement for the proper game. Will the lights then be on or off? Well, ignoring the center, there are 10 LR-spaces, along with 11 L-spaces and 11 R-spaces. So, no matter the order in which the marbles are placed, both buttons will be pushed 21 times, and that means both lights will be *on* when all the marbles are in position.

Now let's play a game of Solitaire. There are only four possible beginning moves, one of which is indicated by the yellow circle above. The only place to move the marble in the circled R-space is to the centre LR-space, and we then remove the marble on the L-space between them. In effect two marbles are removed and one marble is added. Accordingly, we will hit both the R button and the L button twice, and both lights will still be on after our move.

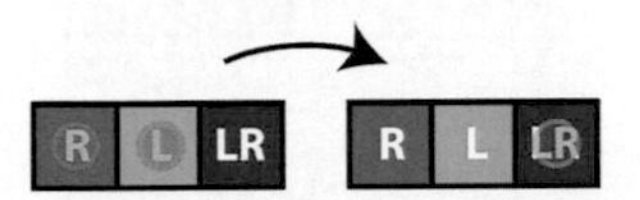

And that demonstrates the trick. The key observation is that any three spaces in a row will include exactly one L, one R and one LR. And, a move consists of adding or removing a marble from each space. It follows that after *any* move at *any* stage of the game, both lights will still be on.

So, however the game ends, it must do so with both lights on. But then, supposing there is only one marble left, where could it be? Removing the last marble will leave an empty board, at which stage we know that both lights must be off. It follows that this last marble must have been on one of the LR-spaces:

That's limited the possible endings and we can now easily limit them further. If we had started with the mirror-image labelling, the exact same argument would have proved that the final marble must be in one of the "mirror" LR-spaces:

It follows that the final marble must actually be in one of the overlap LR-spaces, so one of the five spaces we indicated above. That completes the very pretty proof.

This same clever idea can be used to analyze many variations of Solitaire, with some nice surprises. Suppose, for example, your local Solitaire Master is becoming a little smug. Then you can present him with the 7×7 Solitaire game pictured below. You can make it even easier for him, allowing the final marble to wind up wherever he wishes.

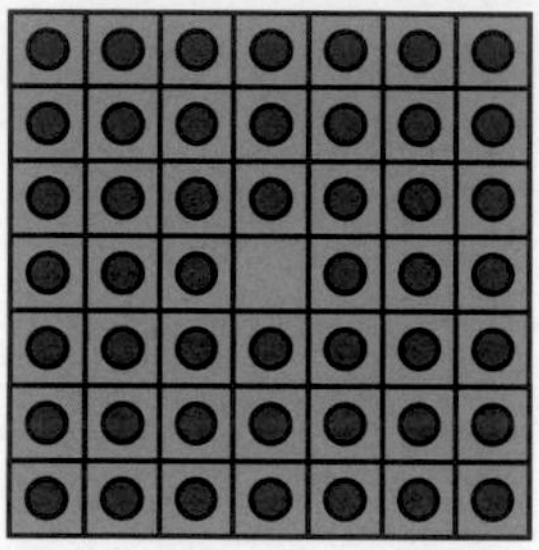

It turns out this new game is impossible and it is now easy for us to prove that this is so. As above, we set up our light gizmo and label the board:

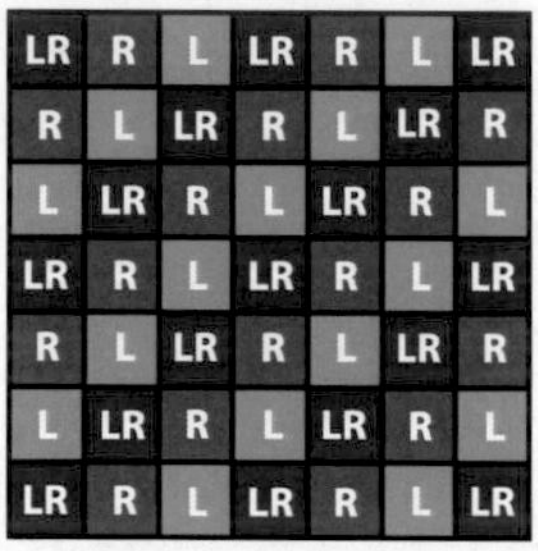

LR	R	L	LR	R	L	LR
R	L	LR	R	L	LR	R
L	LR	R	L	LR	R	L
LR	R	L	LR	R	L	LR
R	L	LR	R	L	LR	R
L	LR	R	L	LR	R	L
LR	R	L	LR	R	L	LR

There will be 16 spaces of each type (not counting the center), which means that if the empty board had both lights off then so will the board at the start of our game. But then with just one marble remaining both lights must still be off, and that's impossible: removing the last marble would result in an empty board with at least one light on.

This is all very cute, but is there any deeper meaning? Not surprisingly, yes.

Our lightbulb gizmo is a simple example of what is known as a *mathematical invariant*: a quantity that remains unchanged when certain transformations are performed. There is much more that invariants can say about Peg Solitaire, and the keen reader might wish to explore further.[1]

Mathematical invariants play very important roles in many branches of mathematics. At times they provide beautifully simple answers to what appear to be fiendishly difficult questions. Alas, at other times the invariants simply fail to help; for example there are Solitaire configurations that our gizmo will fail to rule out but which are in fact impossible to solve.

But that's enough deeper meaning for one column, and it's time for a puzzle.

---

[1]At the time of writing, mathematician Yael Algom Kfir's excellent article *Peg Solitaire and Group Theory* is available online.

Puzzle to ponder

Can you determine whether these two puzzles can or cannot be solved?

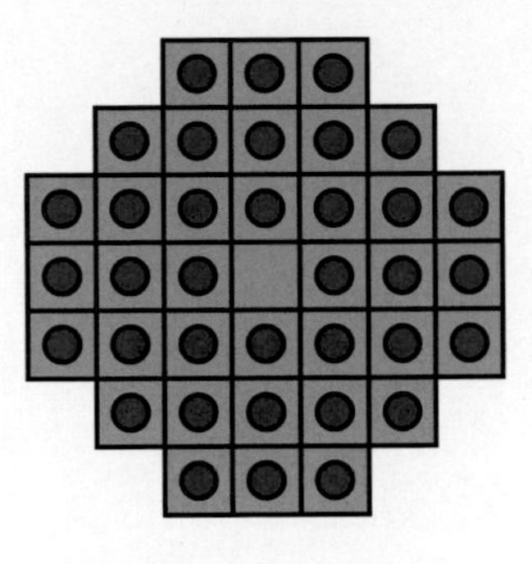

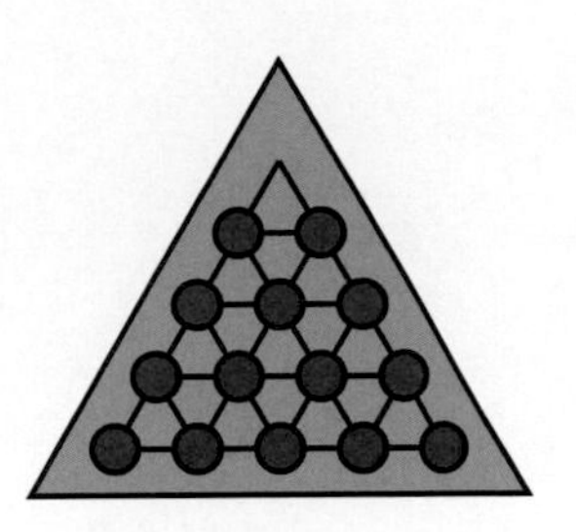

CHAPTER 12

Shadowlands

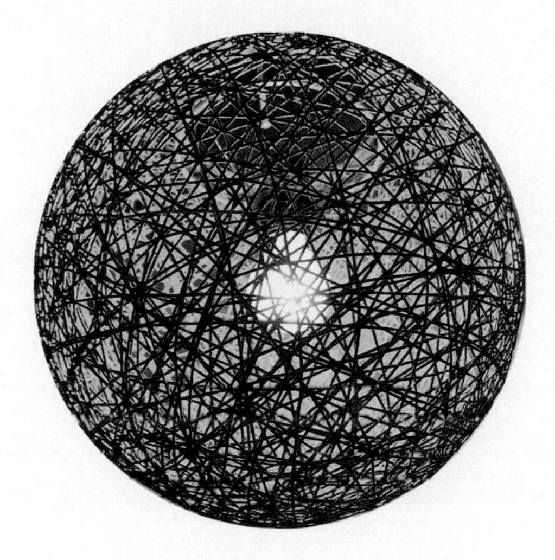

Your Maths Masters recently saved all of Melbourne, raising the alarm when alien spaceships invaded Lygon Street.[1] (You're all very welcome.) Now we're warning of a new threat: large spheres have been seen hovering in the eastern suburbs. Is it time again to head for the hills?

On closer inspection – and we must get into the habit of doing that prior to sounding the sirens – the strange spheres turn out to be restaurant décor. They're very large string lampshades, not from another world, though they are from another time; these massive lampshades were once considered very cool, and they all but conquered the World in the 1970s.

The lampshades are simple to make: just blow up a beachball and wrap it with string soaked in glue. Once the glue has hardened, deflate the beachball and remove it through a small hole.[2] Very easy, if not obviously so mathematical.

It was only with his recent encounter in a Melbourne restaurant that one of your Maths Masters was struck by the mathematics exhibited by the lampshades.

[1] An article we wrote on *Moiré patterns* appearing on outside gas heaters. Lygon Street is very famous in Melbourne, as the center of the traditionally Italian area and, in the 1960s, was the first truly cosmopolitan part of Melbourne.

[2] It is easy to find detailed instructions online.

What is intriguing is that all the shadows cast by the string appear to be straight lines:

Why is that?

The straightness of the shadows can be explained by considering the lampshade's construction. To keep the unset string from sliding off the spherical beachball it is important to wrap the string approximately along "equators", what are known as *great circles.* Mathematically, we can obtain any of these great circles by intersecting the beachball with a plane slicing through the centre. Then, the shadow of that circle on the ceiling will be the intersection of the plane with the ceiling, a straight line.

This arrangement is similar to another, very famous and very important method of shadow making. Instead of placing the light bulb at the centre of our spherical shade, let's position it at the bottom of the sphere. The sphere will now be useless as a shade for people below, but it will cast some impressive ceiling shadows above.

With this arrangement, any special circle on the sphere, one that passes through the bulb, will cast a straight line shadow on the ceiling. This follows from similar reasoning as for our original lamp, and has been beautifully brought to life by our friend and colleague, Henry Segerman.[3] But there is something even more surprising with this arrangement: any circle on the sphere that does *not* pass through the bulb will cast a shadow that is circular.

[3]Henry had an excellent video and many beautiful photographs online, but at the time of writing they no longer seem to be available.

This remarkable property suggests yet another design: construct a lampshade consisting of circles arranged on a sphere. That is exactly what Henry did, using 3D printing, and the result is stunning.

This beautiful, bottom-of-sphere method of casting shadows is known as *stereographic projection.* It was known to the ancient Greeks, and Apollonius of Perga was able to prove the circles-to-circles property we've been admiring. Nowadays, the circle property can be proved either algebraically or geometrically. The proofs are a little too involved to present here, but we can indicate one of the key geometric ideas.

Draw a circle on the ceiling, preferably while the restaurateur has his back turned. Now imagine a cone from the lightbulb to the circle, as pictured in profile below. It is clear that any slice of the cone parallel to the ceiling will also be a circle. What is interesting is that there is another slicing angle that will result in circles.

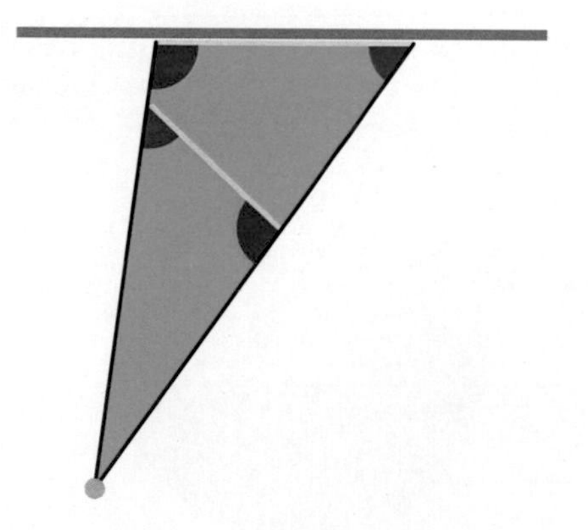

It turns out that if we slice so that the two red angles are equal, implying the two blue angles are also equal, then the cross-section will be a circle. Moreover, such a circle slice will exactly fit on our sphere, and its shadow will be the original circle drawn on the ceiling. Very pretty. (Admittedly, we've omitted the less pretty details.)

Stereographic projection has a second remarkable property. To explain, we'll replace the lampshade by a transparent globe. The light will now project a map of the Earth onto the ceiling:

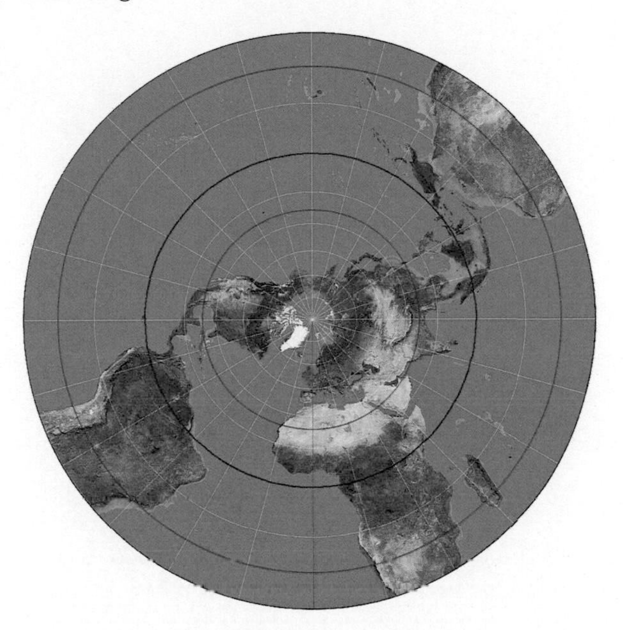

We know that every circle on the Earth will appear as a circle on the map, but something much more important is true: every *angle* on the Earth will appear as the same angle on the map. For example, if two paths along the Earth meet at right angles then the corresponding paths on the map will also meet at right angles. That property of the projection is beautifully illustrated by Henry's lampshade above.

Why is the angle property so important? Because it implies that a compass direction – the angle from North – as pictured on the map, will indicate the true compass direction on the Earth. It is this property that made the similar 1569 map of Gerardus Mercator so revolutionary. In brief, Mercator's was the first world map that reliably indicated which way to go.

So once again your Maths Masters have cried wolf on the alien invasion, but we have been led to some important and beautiful shadow math. And, as we hope to investigate in the future, there are plenty more surprises hidden in the shadows. Just not any aliens. We think.

Puzzle to ponder

Here's an easy one. Can you show that after stereographic projection the circle created by the equator's shadow has an area equal to the area of the sphere?

CHAPTER 13

Picture perfect

There's nothing quite as frustrating as trying to hang a favorite picture and having it simply refuse to hang straight. At least, however, there is a very pretty explanation for this annoying behavior. And, as it happens, there is also a very simple remedy.

Let's consider hanging one of our treasured art works, *Calculator and Hammer*, reproduced above. At first glance it is difficult to see why the print shouldn't happily hang level, as pictured on the left, below. If level, then the two parts of the string will pull symmetrically away from each other and together will balance the downward gravitational pull on the print.

One might just take this as further evidence that calculators skew everything, even gravity. However, this time (and only this time), the calculator is innocent. If the attached string is too short then there is another arrangement for which the forces will balance, resulting in our annoyingly tilted arrangement. And, the underlying reason why the print chooses to be tilted rather than level is indicated by the green dots: the center (more technically, the "center of mass") of the print is lower when tilted.

In the lingo, the problem is that the print is *unstable* when level, so even a very slight shove will start it sliding into the stable, tilted position.[1]

One way to encourage the print to hang straight is to lengthen the attached string. It turns out that the longer the string, the more level the preferred hanging position. Moreover, once the string is sufficiently long the only stable position for the print is to be exactly level.

To determine how much the print will tilt, and how long the string must be to avoid tilting altogether, all that is required is a little classical geometry, a few facts about circles and ellipses. Of course, discussing classical Greek geometry in Australia is akin to *talking* in Greek, so we'll leave most of the details as homework for the aficionados.

Let's consider a balanced position for the print. Then, whether level or not, the centre of the print must lie directly below the picture hook; otherwise, the weight of the print would start the print rotating around the hook. Secondly, in order for the horizontal pull from the two parts of the string to balance, the vertical line connecting the print and the hook must bisect the angle made by the string.

[1] For the details, see F. J. Bloore and H. R. Morton, *Advice in Hanging Pictures,* American Mathematical Monthly, **92,** 309–321, 1985. The article is very interesting but technical. At the time of writing, the authors have a shorter version of the article freely available online.

Now, suppose the print is *not* level. Then we can show that a circle can be drawn through the hook, the centre of the print and the two top corners of the frame.[2]

That's almost all we need. If the string is too long then we simply cannot draw the circle above. It then follows that in order for the print to be stable it must also

[2]This is explained with some care in Bloore and Morton's articles.

be level. It turns out that the critical length depends upon the dimensions of the print, as indicated in the diagram below.

If the string is shorter than this critical length then the print will be tilted when stable. To determine exactly how tilted, we employ one more classical trick. Rather than rotating the print by sliding the string over the picture hook, we'll glue the print to the wall (of our rented apartment) and move the hook around the print. The hook's motion is constrained by the total length of the string, and the path the hook scratches into the wall will be a perfect ellipse. (Our landlord will probably be unimpressed, but this is the famous *gardener's construction* of an ellipse, with the top corners of the frame at the *focal points*.)

To determine how far to slide the hook, we just have to locate where the ellipse intersects the circle that we constructed above:

Finally, we rotate the whole construction to have the centre of the picture directly below the hook, and we're done!

There you have it: some lovely geometry to solve all our picturing woes. Now it's time to add to our gallery. Maybe we'll continue with *Calculator and Bowling Ball.*

Puzzle to ponder

Justify the second picture in the argument above. That is, show that if the print is balanced in a tilted position then the hook, the center of the picture and the two top corners all lie on a circle.

CHAPTER 14

Tractrix and truck tricks

As we have written of time and again, beautiful mathematics is everywhere. And, last Thursday, opening an innocent kitchen cupboard, there it was: some very clever tea bags had formed themselves into a famous mathematical curve, the *tractrix*.

Indeed, tractrices had been stalking your Maths Masters all week. Earlier, while practising some card flourishes and flipping over a ribbon spread, there was another:

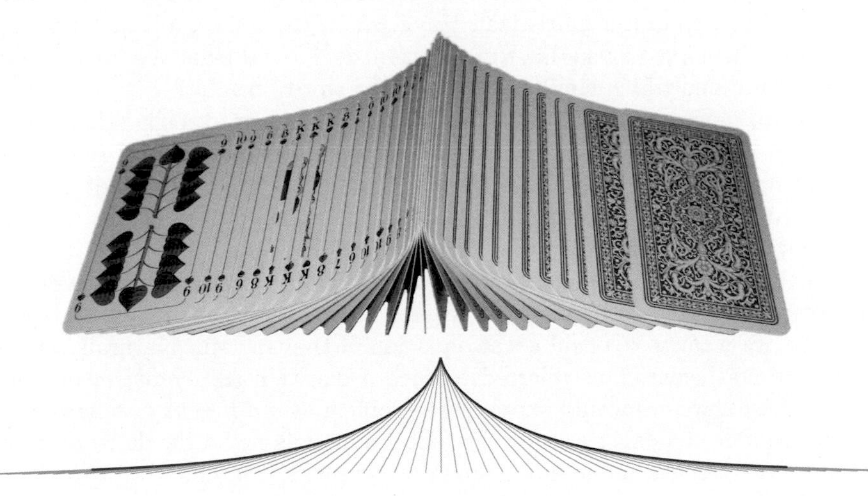

Then, while toppling domino chains with some junior Maths Masters, another tractrix appeared:

Finally, your Masters Masters found themselves caught behind a massive truck. In order to drag its superlong trailer through a T-junction the truck cabin had to begin traversing the corner at right angles to the trailer:

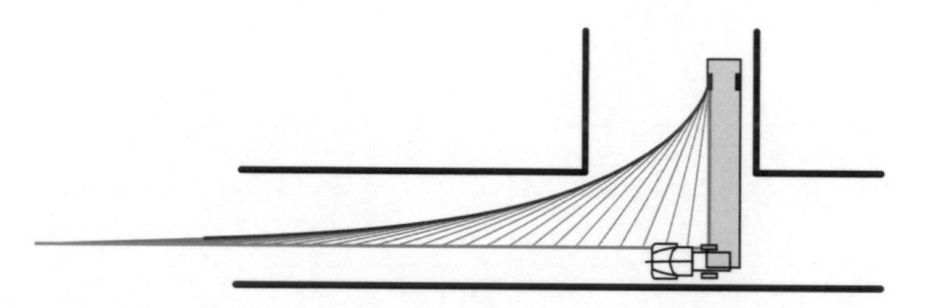

The rear (red) wheel passed through a puddle and traced out a tractrix on the dry street ahead.

The above curves are all striking and they appear to be basically the same. But what exactly makes a tractrix a tractrix?

The curve was first considered by 17th century anatomist and architect Claude Perrault, the much less famous brother of Charles Perrault, the collector-author of *Little Red Riding Hood* and many other classic fairy tales. Claude placed his watch in the middle of a table and pulled the end of the watch chain along the edge of the table. Perrault then asked for the shape of the path traced out by the watch. This is identical to our truck scenario above.

Perrault was not a mathematician and it took the great Sir Isaac Newton and Gottfried Leibniz, and later Christiaan Huygens, to solve the problem. The name "tractrix" for the curve is also due to Leibniz. It is derived from the Latin *trahere*, meaning to pull, and which similarly gives us the word "tractor".

The key to determining a precise rule for the tractrix is to realize that the rear red wheel (or watch) is a fixed distance behind the green front wheel (chain end), *and* that the truck trailer (watch chain) always indicates the direction of motion of the rear red wheel. That is, we know the *tangent* to the tractrix at any point and our problem then is to determine the tractrix itself.

Given the equation of a curve it is easy to determine various properties of that curve, but our traxtrix problem is the reverse of this: given a certain property of the tractrix, we want to determine the equation of the tractrix. Naturally enough, 17th century mathematicians referred to such a question as a *problem of inverse tangents*. These days, we would express the problem as a *differential equation*. The details are a little too calculusy for this book, but it is not difficult to determine and to solve the tractrix differential equation: we sneak it in below as a specialists' puzzle and solution.

Once you have the notion in mind, you'll begin spotting tractrices everywhere. Moreover, the tractrix path and its generalization known as *tractional motion*, where the front wheel needn't travel in a straight line, have become an indispensable tool in the design of street intersections, parking garages and the like.

Suppose that you want to check whether your newly designed roundabout will accommodate a certain size truck. One fun method is to build the roundabout, drive a truck through it, survey the damage and adjust the roundabout accordingly.

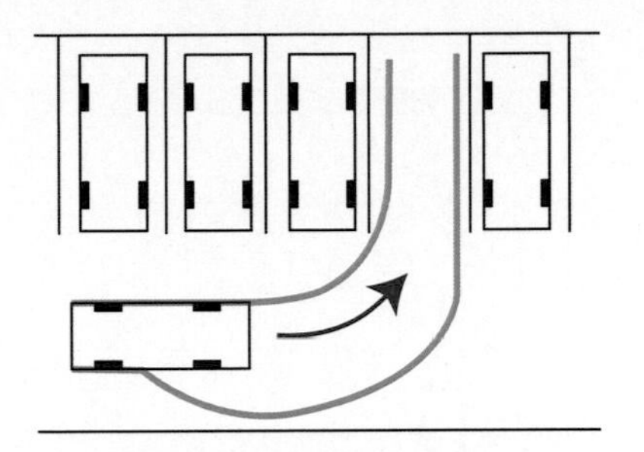

Less fun but a lot cheaper is to perform a *swept path analysis.* As indicated in the above diagram this consists of driving a virtual version of vehicles around your design and calculating the regions swept out by the different components of the vehicles. These swept out regions are typically bordered by tractional curves.

There is much more to the tractrix, and we'll write soon about an astonishing 2D world that the tractrix generates.[1] But for this week we'll stick to the trucks, and we'll close with a simple safety message: doesn't the tractrix demonstrate quite dramatically why you should heed those "do not overtake" signs that are displayed on big trucks?

[1] See the following two Chapters.

Puzzle to ponder

(For calculus fans.) Suppose the distance between the truck wheels is L, that the front wheels start at the origin $(0, 0)$, with the truck lying along the positive x-axis, and then the truck drives along the positive y-axis. Use this scenario to derive the equation of the tractrix.

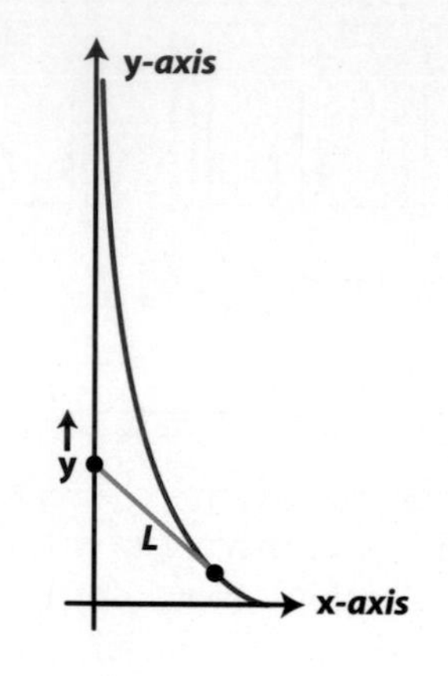

CHAPTER 15

Cycling in circles

As the Tour de France continues,[1] it has been great fun to watch, and only watch, riders on those long, mountainous routes. We have discovered, however, that it might be even more fun if they simply rode in circles.

Take some chalk and draw on the ground a big loop without indentations: what mathematicians call a *closed convex curve.* Now ride your bicycle so that the back wheel follows the path of the loop. As you do so, the front wheel traces another, larger loop. Together, the two loops bound a ring. Now here's the really cool thing: *regardless of the size and shape of the loops, the area of the ring will always be the same.*

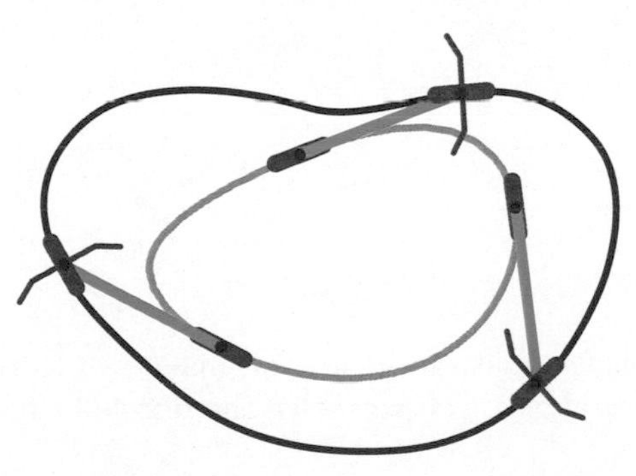

In fact, if your bike is length L from wheel hub to wheel hub, then the area of the ring is always πL^2. That's very cool, and that very familiar quantity suggests something of what's going on.

[1]See Chapter 30.

Imagine the original "loop" is simply a point, and so the back wheel remains stationary. Then the loop traced out by the front wheel is exactly a circle of radius L, and our formula simply gives the area of this circle.

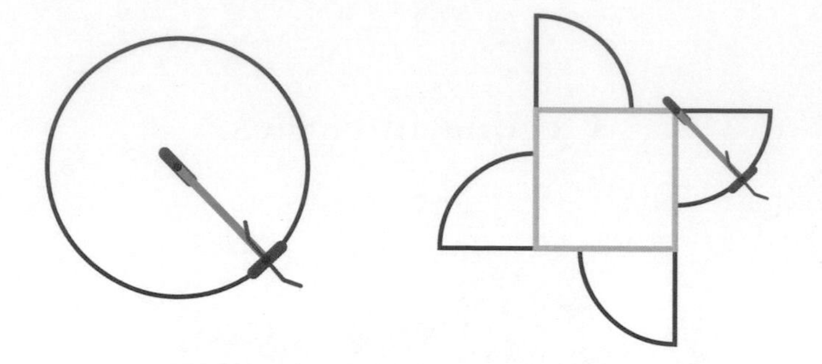

What if we start with a genuine, not-dot loop? As a second example, imagine that the inner loop is a square. Then it is not hard to see that the outer loop traces out four extra quarter circles, and the area formula then follows again. In fact, exactly the same argument works when the inner loop is any convex polygon. Finally, for a genuine, curvy loop, the formula follows by approximating the loop by polygons. Easy.

You can also check that in both of the examples above the *length* of the outer loop is $2\pi L$ greater than the length of the inner loop, and that this formula is also true for convex polygons. So, maybe that's another familiar formula that is always true? Nope. The length formula definitely fails whenever the inner loop is genuinely curvy. We'll leave it for you to ponder why approximating by polygons doesn't work here.

All of this is reminiscent of wheelchair racing champion Kurt Fearnley, when we had Kurt ride around in loops.[2] If Kurt's wheelchair is of width L then we found that the length of the outer loop will indeed always be $2\pi L$ longer than the inner loop. In this scenario, however, the area formula is no longer true.

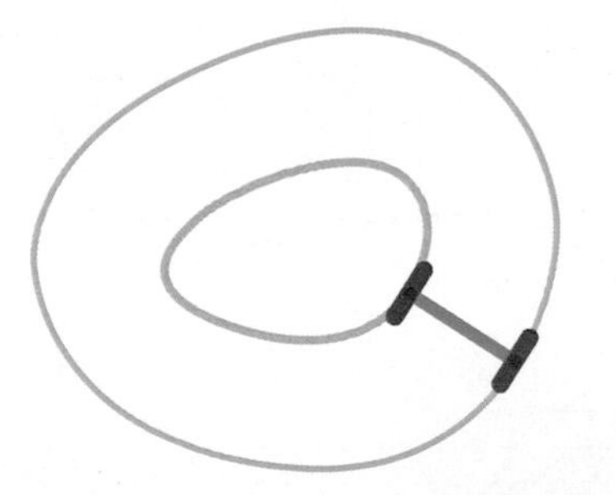

And now, with bicycle math in hand, you may wish to revisit a challenge we set earlier.[3] See if, at a glance, you can solve the puzzle below.

[2]Kurt Fearnley was a very successful Australian wheelchair racer. One of our earliest columns, looking at the mathematics of wheelchair paths, was framed around Kurt.

[3]It became a tradition for our final *Age* newspaper column of the year to be a Summer Challenge: thirty puzzles arranged into categories of Easy, Medium and Hard. The two-ring puzzle on the next page was one of these.

Puzzles to ponder

Which of the two rings below has the greater area?

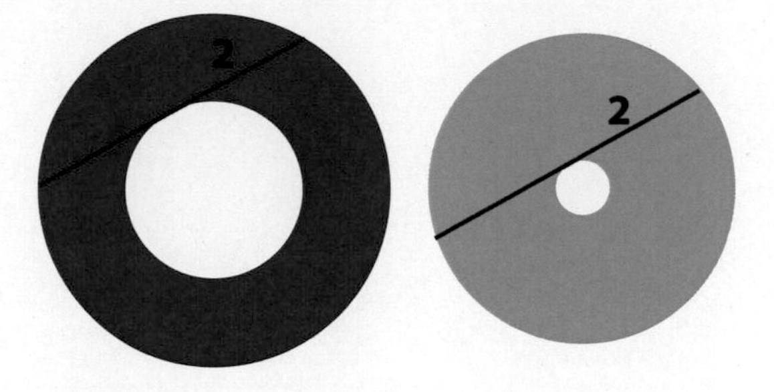

CHAPTER 16

Which way did Natalie go?

The Tour de France is well under way and, much to our surprise, this year's route seems to have taken in the Clayton campus of Melbourne's Monash University.[1] The compelling evidence is in the photo above, showing tire tracks in the wet cement.

The junior maths masters in the foreground are trying to get to the bottom of this. With brilliant detective work, they have already deduced that the stray cyclist is someone named Natalie. They deduced this from viewing Natalie's signing of her masterpiece in the lower left corner.

Our detectives now want to figure out which track was made by the front wheel and which by the back wheel. And, they want to determine in which direction Natalie was riding. In fact, this is a very old puzzle: Sherlock Holmes and his

[1]See Chapter 30.

sidekick Watson puzzled over the very same problem in *The Adventure of the Priory School*:

"This track, as you perceive, was made by a rider who was going from the direction of the school."

"Or towards it?"

"No, no, my dear Watson. The more deeply sunk impression is, of course, the hind wheel, upon which the weight rests. You perceive several places where it has passed across and obliterated the more shallow mark of the front one. It was undoubtedly heading away from the school."

Holmes is unquestionably a great detective, but in this instance he is, well, off track. Regardless of the direction of travel, if there is a crossing then it is the back track crossing over the front track. So, all that you are able to deduce is which track was made by which wheel, and not, as suggested, the direction of travel.

Can we do better than Sherlock Holmes? The following is an aerial view of a bicycle leaving a pair of tracks as it travels from left to right.

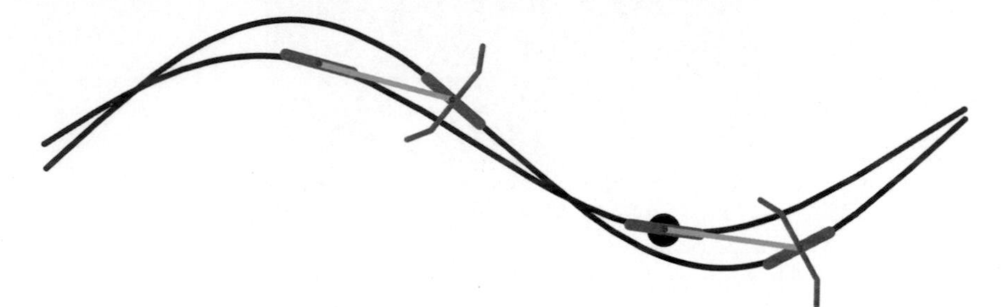

Notice that the back wheel is always pointing towards the middle of the front wheel, and remains a fixed distance behind. A second crucial observation is that the back wheel is always *tangent* to its track, pointing in the direction of its motion. Finally, the back track is generally less curvy than the front track.

Using these observations it is usually very easy to solve bicycle track puzzles, even without Holmes's insight into the back track crossing on top of the front track. For example, in the following picture track 1 is distinctly less wavy, which suggests that it is the back track.

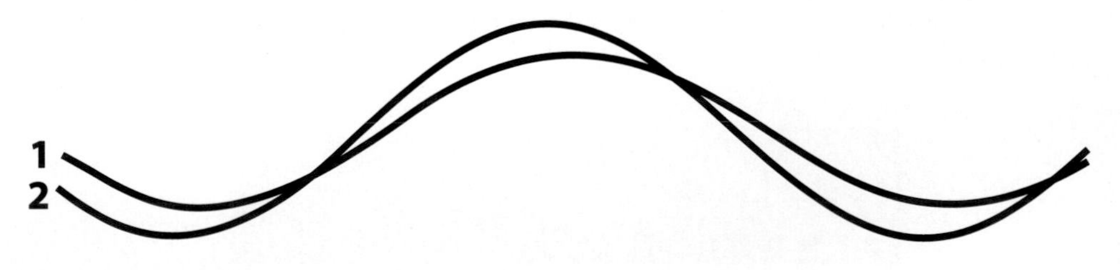

Next, we can draw some tangents to this suspected back track and see where they intersect the front track.

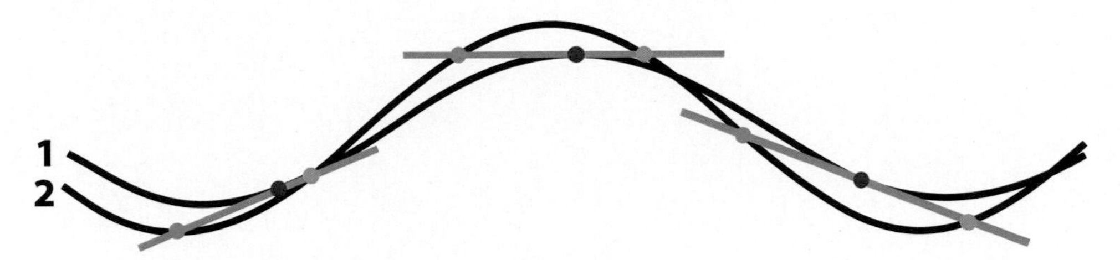

If the bicycle were travelling to the right, the yellow point would mark the corresponding position of the front wheel; if travelling to the left, the green point indicates the front wheel. But the distance between the yellow point and brown point keeps changing, which is impossible. By contrast, the distance between the green point and the red point is always the same. So, we can conclude that the green point correctly indicates the front wheel, and therefore the bicycle was travelling to the left.

This doesn't always work. If you ride your bicycle in a circle then the two tracks will be concentric circles, and the circles will be the same no matter the direction in which you ride. Or, if you ride in a perfectly straight line then you'll see just one track. And, surprisingly, there are other bicycle tracks that can be traversed in either direction, and other "unicycle tracks" where the back wheel exactly follows the front. But these weird tracks are pretty much impossible to ride: in practice, our detective procedure works very well.

Now, back to Natalie's tracks. Below we have reproduced a bird's eye view of her tracks, and you should have no trouble determining Natalie's direction. Or, if you happen to spot Natalie at the finish of the Tour, then do please ask her for us.

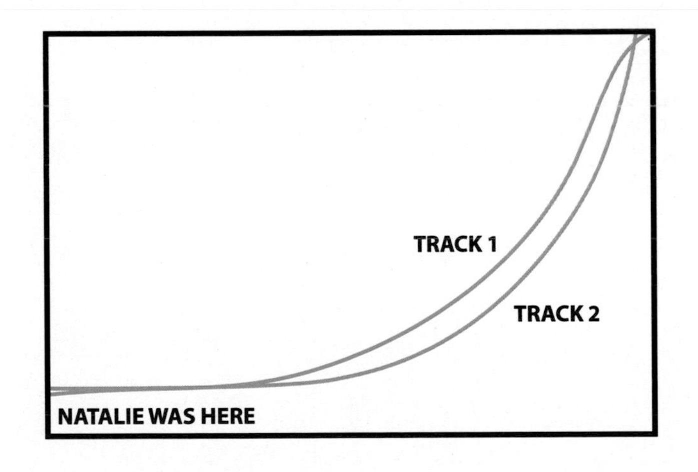

Puzzle to ponder

Which way did Natalie go?

CHAPTER 17

$\pi = 3$

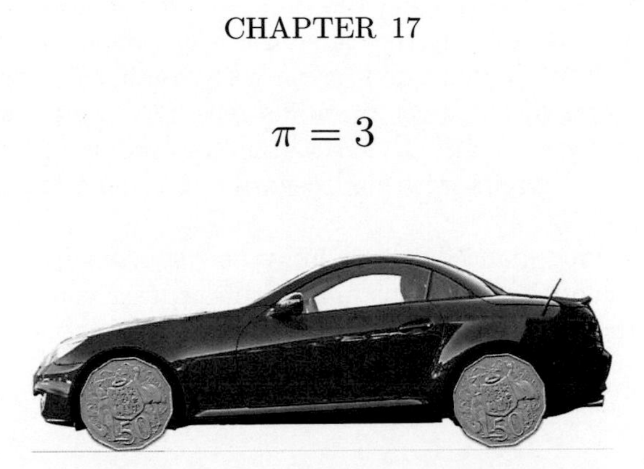

Did you all enjoy March 14? That was Pi Day, when we celebrate the mathematical constant π. Why March 14? That's because the Americans who came up with the idea write the date as 3.14.[1] Of course 3.14 is not exactly π, just a convenient approximation. But if we really want usefulness together with ease, nothing betters 3 as an approximation to π.

It would be so much more convenient if π were actually 3. For those who take the Bible literally, this is already true. In *Kings* 7:23 we read of the dimensions of a sea (a type of vessel):

He made the Sea of cast metal, circular in shape, measuring ten cubits from rim to rim and five cubits high. It took a line of thirty cubits to measure around it.

Dividing the circumference of 30 cubits by the diameter of 10 cubits, we conclude that $\pi = 3$.[2]

That is just silliness of course, but here is something true, and quite amazing. Have a look at the following two diagrams.

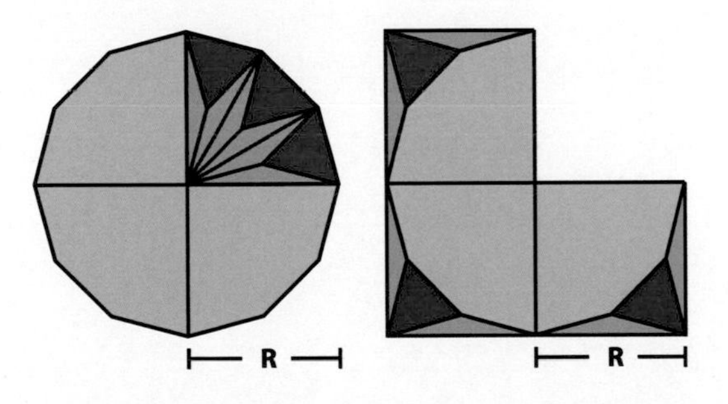

[1] For a while, Australians celebrated July 22 as Pi Day, since Australians write that date as 22/7, with the month second. The Americanization has won out, however, and March 14 it is.

[2] Without diving into deep apologetics, it seems fair enough to take the passage as intended to be interpreted in an approximate manner, for what would have been a bloody big pot. There's no evidence that any Christian fundamentalist ever took the passage to mean God was commanding that π be 3.

The first diagram is a regular 12-sided polygon, known as a dodecagon – the shape of a 50-cent coin. The second consists of three squares of sidelength R.

With a little cutting and pasting, you can transform the dodecagon into the three squares, simply by rearranging the blue and red triangles. It follows that the two shapes must have exactly the same area. Of course, the area of the three squares is $3R^2$. So, the area of the dodecagon with "radius" R is also exactly $3R^2$.

Now, since the area of a "circle" of radius R is πR^2, we know exactly how to arrange things so that π is exactly 3. Our Maths Masters plan for a better universe is simply to replace all circles with dodecagons. Mathematical life would be so much easier.

But perhaps, having glanced at our shiny new sports car, you are somewhat sceptical of the value of this ingenious plan. Still, at least you now know the easy way to calculate the area of a 50-cent coin.

Puzzles to ponder

What if we focus upon perimeter rather than area? Suppose we wish for the "circumference" of our "circles" to be $2\pi R$, and then try to arrange for π to equal 3. Is there a shape that will do this for us?

CHAPTER 18

Just the right level of wine

Your Maths Masters have been drinking quite a lot of wine of late. Never minding our motivation (it's Tony Abbott),[1] we've discovered something very interesting about bottle-stacking.

It was by accident. Our bottles had been stacked simply and symmetrically, just as one usually sees:

But then some junior maths masters decided to reorganize things. The result? Many broken bottles, two grumpy Maths Masters and some hasty and haphazard restacking.

[1]Australia's reactionary Prime Minister at the time. Abbott's successors have been microscopically better.

These junior bottle-stacking efforts didn't begin auspiciously:

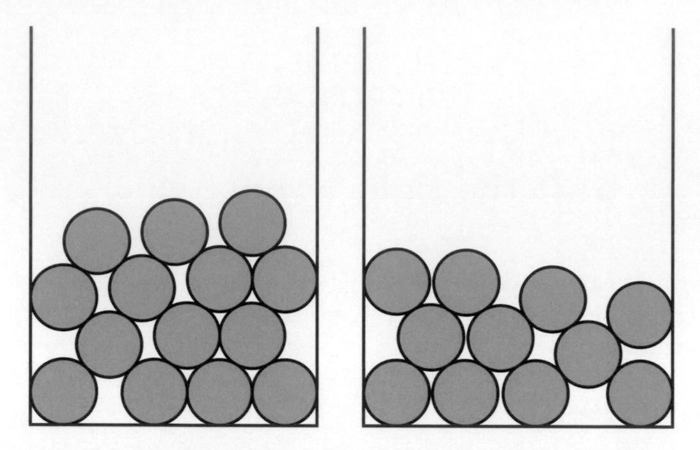

After a little more stacking, however, something quite amazing happened. The seventh rows worked out to be completely level.

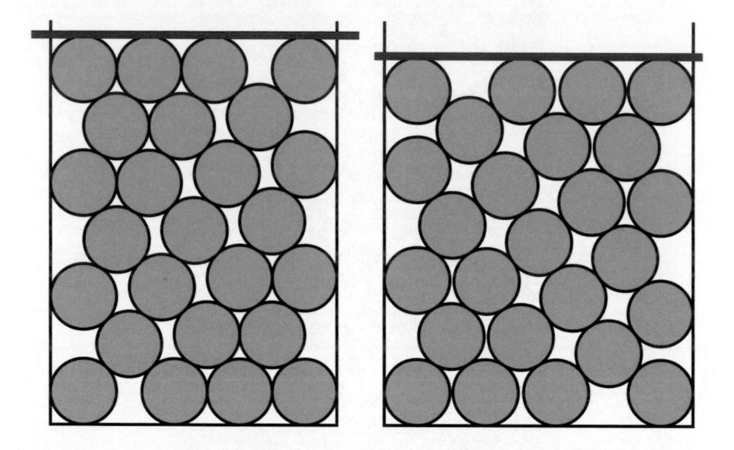

It turns out that something like this will (almost) always occur, a discovery of French mathematician, Charles Payan. If our wine closet is wide enough to accommodate four bottles along a row, and if we begin with two bottles in the bottom left and right corners then, pretty much no matter how the bottom row is completed, the seventh row will wind up level. This can only fail if the closet is a bit too wide, almost enough to permit five bottles in a row. Similarly, if the closet is wide enough for five bottles along a row then (almost) any crazy stacking will be level along the ninth row. For a six-bottle wide closet we need to go up to eleven rows, and so on.

These stacks have other surprising properties. In particular, the middle bottle of the middle row will always be exactly halfway between the two walls. And, a wine stack will always have a 180-degree symmetry around this central bottle.

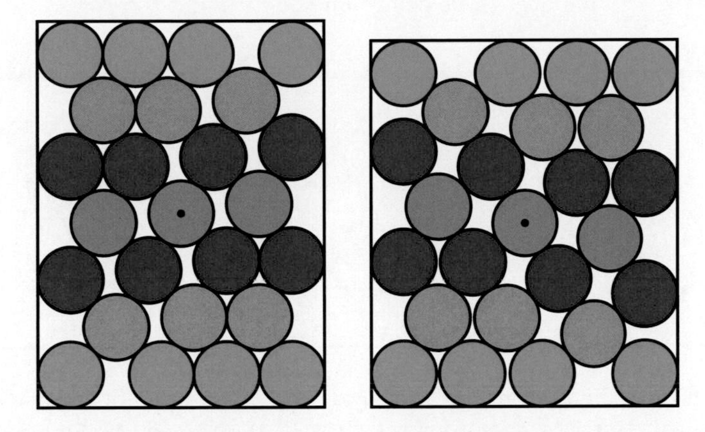

This provides us with one of the weirdest and silliest methods of finding the midpoint(s) between two parallel walls. Just stack a wonky pyramid of bottles between the walls. Then the top bottle of the pyramid will be our "centre bottle", and so will be an equal distance from the two walls.

The superb Cut-the-Knot website has some nifty java applets (and some nitty-gritty proofs) which can be used to explore these strange stacks.[2] But what underlies it all?

This stacking phenomenon can be explained with just a few rhombuses and isosceles triangles. Suppose three circles of the same size form a little open necklace, as pictured below. Then joining the circle centres creates an isosceles triangle. If we then include a fourth circle to complete a necklace the centres form a rhombus: all sides are of equal length and opposite sides are parallel.

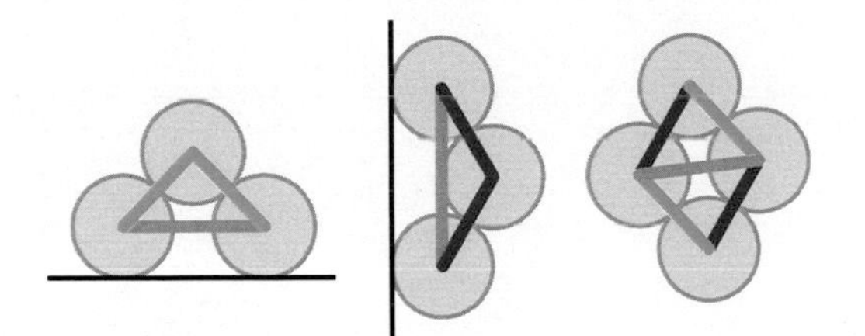

[2]Cut-the-knot was started by the mathematician Alexander Bogomolny, who died in 2018. At the time of writing, this extraordinary online resource is still being maintained and is still accessible.

To see where this fits in, let's consider a smaller closet, just wide enough to accommodate three bottles along a row, as pictured below. Starting at the centre of any circle on the bottom row, we can step our way up along the edges of triangles and rhombuses, arriving at a circle centre on the fifth and top row.

Any path from the bottom row to the top row will consist of two red edges and two green edges. But it is easy to check that all the red edges climb at the same angle, and similarly for all the green edges. So, all the circle centres in the fifth row must be at the same height. Done!

For a wider closet, with more bottles along the bottom row, there are more fiddly details, with more differently angled edges. However, the same triangle and rhombus approach still works.

We'll close by making one last surprising and important observation. If one of our special wine stacks is rotated by 90° then the result is another special stack. That's very handy to know, just in case a mischievous little maths master decides to tip over our precious wine closet.

Puzzle to ponder

Show why a wide 3-bottle closet may not work.

CHAPTER 19

Spotting an unfortunate spot

If for no other reason, 2011 will be remembered for its tragic earthquakes: first Burma, then China and Christchurch, and then the truly disastrous earthquake in Japan.

Earthquakes affect very large regions. To pinpoint the exact source of an earthquake, seismologists employ a world-wide network of seismographs, which are continuously registering ground movement. And, there is some very clever mathematics.

Above is a typical seismographic record of an earthquake. More details can be found online, but in brief the seismograph has recorded two types of waves emitted from the source, the *focus* of the earthquake: there are fast-moving primary waves and slower secondary waves. The seismograph records the amplitude of the waves and the times that the waves are detected.

Seismologists also know the speeds of the primary and secondary waves. So, the time interval between the detection of these waves can be used to determine the distance of the seismograph from the earthquake's focus.

Now, if your seismograph indicates that there has been an earthquake 100 km away, this means that the focus of the earthquake is located somewhere on a circle of radius 100 km, with your seismograph at the centre. However, to determine exactly where the focus is on this circle, you'll also require the circles obtained from other seismographs.

Two circles corresponding to the same earthquake will intersect at either one or two points. In either case, the circle from a third seismograph will be able to locate the focus of the earthquake.

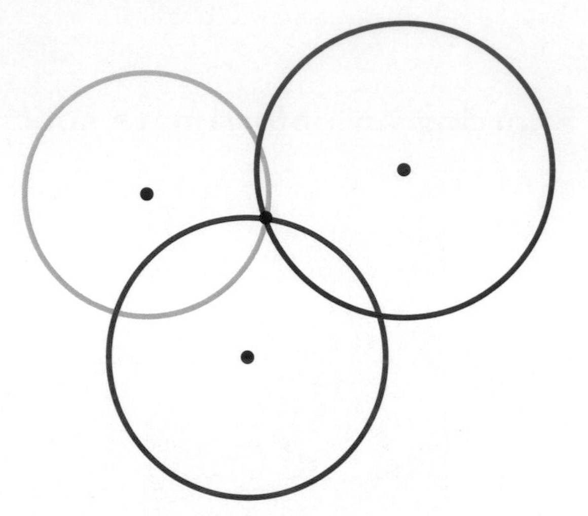

Of course, this description oversimplifies the many details. For one thing, earthquakes do not always occur close to the Earth's surface. This means that instead of drawing circles around each seismograph, we should be imagining spheres. Still, the same ideas work. Two spheres will either intersect in a single point, which must then be the focus, or in a circle. In the latter case, a suitable third sphere will intersect the circle in two points. One of these points may be above ground level, which is an unlikely spot for an earthquake, and so the other point must be the focus. In any case, a suitable fourth sphere will always suffice to locate the focus of the earthquake.

Similar calculations play a role when a location has to be determined from the distances from certain measuring sites. For example, the worldwide GPS system relies on a system of satellites orbiting the Earth. The satellites carry synchronized clocks that are constantly sending out signals, indicating the time when the signal is sent, and the coordinates of the satellite at that moment. Your GPS device then combines the information carried by the signals from (at least) four different satellites, to calculate your position.

By the way, news reports on earthquakes typically contain three mathematical pieces of information: the *epicenter*, which is the point on the Earth's surface directly above the focus; the depth of the focus; and the magnitude of the earthquake. The epicenter and the depth of the focus are easily calculated once we know the location of the focus. Finally, the magnitude can be calculated from the amplitude of the seismograph output, together with the distance of the quake from the seismograph.

Puzzle to ponder

Let's consider an earthquake with focus on the Earth's surface. So, we can stick to drawing circles rather than spheres. Can you think of how three seismographs might be located so that the focus of this earthquake *cannot* be pinpointed?

CHAPTER 20

Summing up the Mystery of Flight MH370

Your Maths Masters usually endeavor to make these columns light-hearted and funny, but definitely not this week; there is nothing remotely light-hearted about the disappearance of Malaysian Flight MH370. As did many people, your Maths Masters followed intently the desperate and (so far) unavailing search. We want to explain some of the mathematical detective work directing that search.

Some important technical information has not been released. However John Zweck, an Australian mathematician based at the University of Texas at Dallas, and a good friend of your Maths Masters, has worked hard to connect the publicly available dots. We'll try to explain the ideas behind John's analysis.[1]

MH370's transponder ceased contact about one hour into the flight. Military radar then tracked the plane's path. The problem was to determine the path of the plane after the cessation of radar contact.

Even after vanishing from radar, MH370 continued to respond to hourly "pings" from Satellite INMARSAT 3F-1, the responses ending around the time the plane would have run out of fuel. The probable flight path had to be inferred from those pings.

[1]John has provided us with extensive help on this column. At the time of writing, many more details can be found at the links on John's university homepage.

In the diagram below the black dot represents the approximate location of the plane when radar contact was lost. The purple cross represents the spot on the Earth directly below 3F-1; the satellite is *geostationary*, which means that the cross (pretty much) stays in the same spot above the Earth.

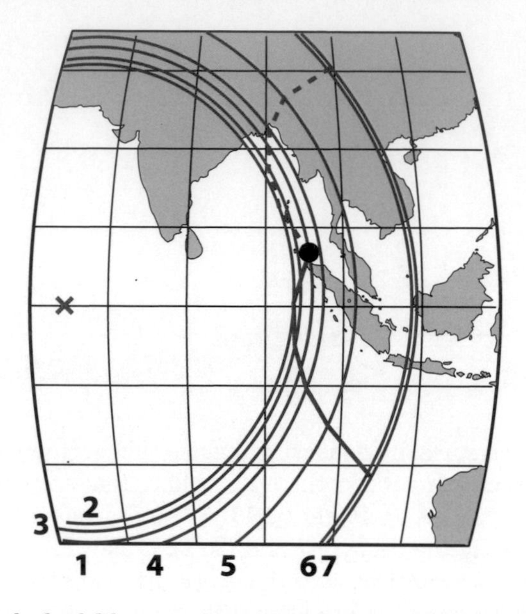

The solid and dashed blue curves represent the two possible flight paths, one northerly and one southerly. We'll try to explain how those paths were calculated, and how it was determined that the plane flew south rather than north.

Each time 3F-1 pinged it took a certain period of time for the reply to return from MH370. From the length of that time interval it was straight-forward to infer the distance of the plane from the satellite. It followed that the plane at that time must be located somewhere on a circle on the Earth's surface, with the purple cross at the circle's centre.

The seven red circles in the diagram indicate the location circles for MH370 at the hourly ping times. So, whatever the path of MH370, it had to cross all of the red circles at just the right times. Still, that obviously leaves a huge range of possible paths. And from here the analysis becomes much trickier.

The essential clue was provided by the *Doppler shift*. This is the phenomenon that causes the siren of an approaching ambulance to sound at a higher pitch than that of a stationary ambulance; moreover, the increase in pitch enables one to calculate the speed of the moving ambulance. A similar but much more intricate Doppler analysis of the plane's signal, combined with the spacings of the red circles, permitted the experts to determine the speed of the plane at the moment of each ping, as well as the angle the plane's path made with the red circle.

Now, with a suitable model of MH370's flight, we can go about reconstructing its path. Let's begin by considering the path from the starting point on the first circle until it reaches the second circle. We know the angle the plane's path makes with the first circle, which means there are two possible directions the plane may be heading, as indicated by the shaded grey angles in the picture below. The resulting (approximate) paths are indicated by the solid and dashed blue lines.

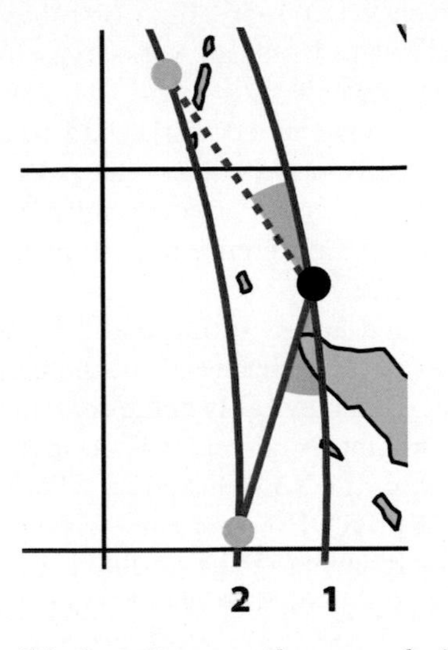

That gives us two possible locations on the second circle. From each location we can then perform a similar calculation, to recreate the two possible paths to the third circle. (Theoretically there are actually four paths to consider, since there are two choices of direction at each of the two points on the second circle. However the paths we have not pictured would include a dramatic change in direction, and so are not plausible as the route of MH370.)

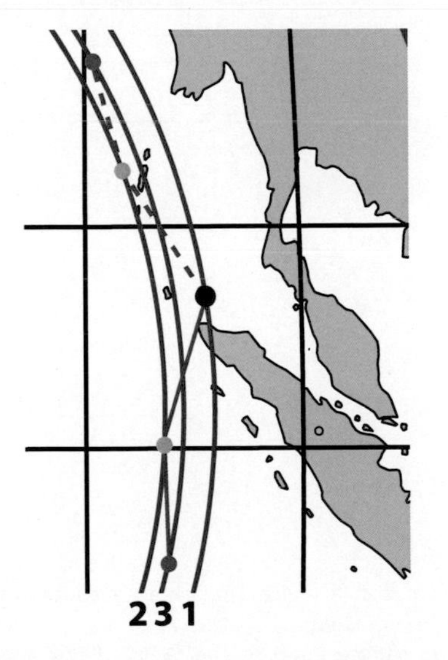

We continue in the same manner until the paths have extended to the seventh circle. The crash site should then be close to the final point on one our two curves. But which one?

The curves are basically heading north and south, and it turns out the experts could rule out the possibility of a northern route. Satellite 3F-1 is not precisely geostationary, meaning the purple cross moves around a bit. This creates a slight asymmetry in the way signals are received, and again the Doppler shift comes into play: the asymmetry allows the experts to distinguish between signals from the north and from the south.

So, that's where we've ended up searching: a remote part of the Indian Ocean, about 1850 km west of Perth. But if this all works, why has the search area changed a number of times, and why haven't they found the plane?

The sad reality is, a few pings is very little data to go on, and the method we have described can only roughly approximate the path of MH370.[2] The approximation is also very sensitive to error; a small shift in the starting point of the path (the black dot) can lead to a difference of hundreds of kilometers in the path's endpoint, the final search region.

As the experts have gained access to more and better data, and as they have refined their analysis of that data, their guesses of the flight path have become more refined and more confident. But inevitably there remains a vast and remote region to search, in ocean that is kilometers deep. It is an incredibly difficult search.

Will MH370 ever be found? No one knows. The search will likely be long and expensive, and could still fail. However some clever and fundamentally simple mathematics has dramatically changed the outlook: a vanishingly small chance of finding MH370 has become a difficult but definite hope.[3]

---

[2]It is essentially *Euler's method*, a technique for approximating the solution of a differential equation, familiar to most senior mathematics students.

[3]Six years later, that hope has dimmed to the point of non-existence. In 2015, about a year after the flight's disappearance, debris positively identified as being from MH370 was found off the East coast of Africa. The location of this debris was claimed to be consistent with the debris having drifted from the general search area, but it is a huge and incredibly remote area.

CHAPTER 21

Choc-full of mathematics

What can be more delightful than chocolate combined with mathematics? The answer, of course, is chocolate combined with more chocolate. Nonetheless, Austria's Mozartkugel is a very tasty mathematical treat.

The Mozartkugel is spherical, with layers of marzipan and nougat and the like, all coated in chocolate. The delicious specimen pictured above is from the Mirabell company but there are many manufacturers.

Which Mozartkugel is best? Ever willing to sacrifice themselves in the name of research, your Maths Masters conducted a comprehensive survey. Happily, it turns out they're all very good, but we've declared Mirabell the winner: as they love to boast, only Mirabell's Mozartkugel is truly spherical.

Having made our choice, and sampling one more Mozartkugel (just to make sure), we can begin to ponder the foil wrapper: how exactly does one wrap a spherical sweet? One way or another, it gets a little clumsy.

The Mirabell wrapper, pictured below, is a longish rectangle. To wrap the Mozartkugel we first form the wrapper into a cylinder, with the red line hugging the chocolate sphere along its equator. Then we just make it work, scrunching things together at the top and bottom.

Other manufacturers use square wrappers. In this case we begin with the north pole of the chocolate sphere touching the centre of the square, and then we scrunch things together at the south pole.

Now, an interesting question, especially for a manufacturer concerned to lower costs: which has less area, the rectangular wrapper or the square wrapper? It's a nice puzzle to ponder, so we've hidden the answer in the following paragraph.

Assuming it *just* covers, the length of the rectangular wrapper must equal the circumference of the Mozartkugel, and its height must be half this circumference. So, if the Mozartkugel has a radius R then the area of the rectangular wrapper is $2\pi R \times \pi R = 2\pi^2 R^2$. Similarly, for the square wrapper, the diagonals must equal the circumference. Then, with a little help from Pythagoras, it is not difficult to show that the square wrapper also has an area of $2\pi^2 R^2$: exactly the same!

The perhaps surprising conclusion is that costwise it doesn't matter whether we use square wrappers or the longer rectangular wrappers. But, we can do better than both.

Since rectangles fit together seamlessly there is no waste when cutting them from a large sheet of foil. Rather, the waste comes from the scrunching, and it turns out the square and rectangle have about 1.6 times the area of the sphere; about 36% of the foil gets scrunched. (We'll leave the true penny-pinchers to determine the precise figures.)

One alternative is to begin with a very long, very thin rectangle and wrap it around and around the sphere, as if pasting the peel back onto an apple. In this way we can lower the scrunch percentage to as close to 0 as desired. But of course such a wrapping would be entirely impractical, and it'd be pretty much impossible to include the pretty picture of Mozart. It turns out, however, that there are other, genuinely practical ways to significantly reduce scrunching.

The simple and clever idea is to cut away chunks of foil that will only be scrunched anyway. For example, beginning with the Mirabell wrapper we can cut away curvy triangles, leaving four petals that will still wrap the sphere. (Renaissance cartographers used similar petal maps of the Earth in the construction of globes; for their purposes, small gaps, mostly around the Poles, were preferable to overlaps.)

It's a tricky question, however, determining petal shapes that will work and are efficient. For example, choosing circular arcs for the edges will result in the petals being a little too skinny, leaving gaps in the wrapping. We'll give some suggestion of why this is true in the puzzle solutions.

Such petal wrappings were investigated in an elegant and fun (and technical) paper by mathematicians Erik Demaine, Martin Demaine, John Iacono and Stefan Langerman.[1] They were inspired by Mozartkugel wrappings to determine the most efficient petal wrappings; in turn, their lovely paper inspired this column.

Demaine and his fellow petalmeisters determined the optimal petal shape to minimize scrunching. The final formula isn't so pretty, but the pretty petals are as pictured above. And, the scrunching percentage for this best petal wrapping is an impressively low 7%. (That can be further lowered by including more and thinner petals, but we would again lose our picture of Mozart.)

Of course, even if the scrunching percentage is low, we'll still be wasting, or at least be forced to recycle, the removed curvy triangles. However, the Demaine and co note that the petal wrappers can actually be very tightly packed, resulting in very little discarded foil:

[1] *Wrapping Spheres With Paper*, Computational Geometry, **42**, 748–757, 2009.

And, we can make everything much neater by expanding each wrapper into a slightly longish hexagon, as pictured below. Now we just have to cut in straight lines, no foil is discarded and the scrunching percentage is still a respectably low 13%.

There are tons of other interesting material in the petalmeisters' paper. For example, they also observe that the same four petals can be used to replace the square wrapper:

Unfortunately these four-petal wrappers cannot be packed so efficiently. However, the companion three-petal wrappers pack extremely well, and they have a number of other desirable properties:

There's plenty more. And all of it inspired by one delicious mathematical sweet.

Puzzle to ponder

Suppose your Mozartkugel has radius R. What are the radii of the circular arcs in the (failed) four-petal wrapper?

CHAPTER 22

Too hot, too cold, just right

Once Upon a Time, Goldilocks went to the cottage of the One Bear. She found a single bowl of porridge, but it was too hot. Goldilocks decided to wait, but she knew that if she waited too long, the porridge would be too cold. At just the right moment, however, the porridge was just right and she ate happily ever after. (Or, at least, until the bear came home.)

This simple idea is captured by the mathematical notion of *continuity*: the temperature of the porridge cools progressively, with no dramatic jumps. The *intermediate value theorem* (IVT) then tells us that if the temperature begins higher than desired and ends up lower than desired then, at some intermediate time – although we don't know when – the temperature will be just right.

IVT is very intuitive, although it is difficult to prove. IVT also has some surprising consequences. Returning to Goldilocks, for example, we find her at the Bear's table, trying to eat her Just Right porridge. However, the floor of the cottage is uneven – bears are not great carpenters and so the table is wobbling. As one often does, Goldilocks looks for a magazine or a piece of paper to jam under one leg. Alas – bears are also not great readers – Goldilocks cannot find any paper. What is poor Goldilocks to do?

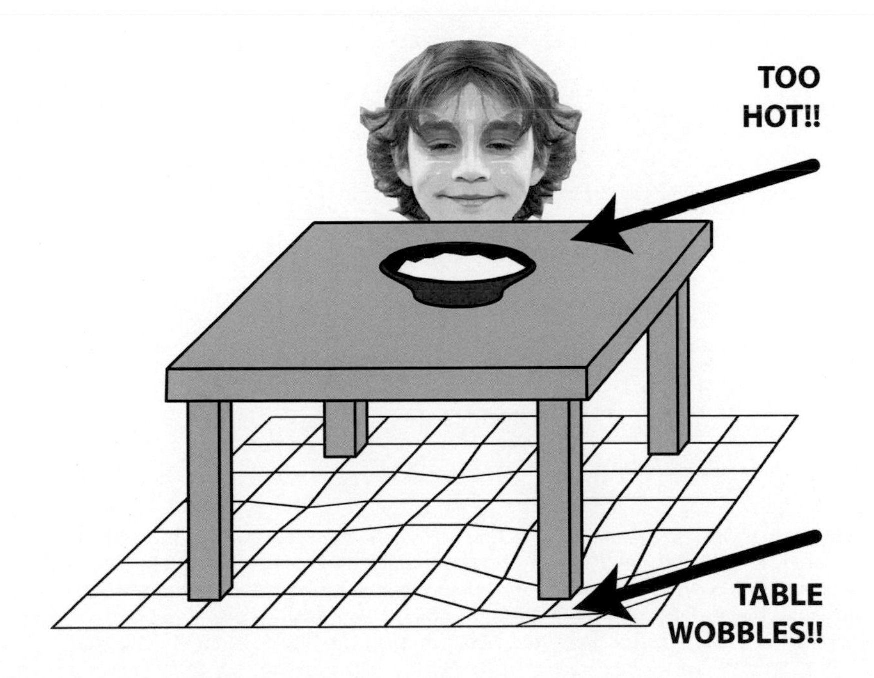

It is here that IVT comes to Goldilocks' rescue. In fact, the table can be balanced, although perhaps still a little tilted, simply by rotating it on the spot. How can we see this?

To begin, balance three of the table legs on the ground, with the fourth leg hovering in the air. Then, try to rotate the table a quarter-turn, keeping the same three legs touching the ground. In fact, although this is not so easy to see, completing such a 90° turn would result in the fourth leg digging into the ground. So, the leg started up too high, and after rotating 90°, it would end up too low. So, by IVT, there is some intermediate angle where the fourth leg is just touching: the table is then balanced, and Goldilocks can again eat happily ever after, for a while.

Disclaimer: The arguments above contain some technical assumptions and details, which we'll explain to Goldilocks once she goes to university.[1]

Puzzle to ponder

The Bear returns, and begins chasing Goldilocks around and around his house, in a big circle. The temperature along the circle is different at different points (but doesn't change with time at a given point). Prove that at some time, Goldilocks and the Bear are at the exact same temperature.

[1] The table turning argument goes back to a *Mathematical Games* column by Martin Gardner, and appeared in his collection *Knotted Dougnuts and Other Mathematical Entertainments*, W. H. Freeman, 1986. Nailing down a proof of the argument turns out to be quite tricky. Together with our colleagues Bill Baritompa and Rainer Löwen, we provide a proof and lengthy discussion in *Mathematical Table Turning Revisited*, Mathematical Intelligencer, **29**, 49–58, 2007.

Part 3

A Bloody Good Sport

CHAPTER 23

Seeds of doubt

It's tennis time. The French Open is in full swing and we're all anticipating a terrific Djokovic-Nadal final. True, Roger Federer may upset that plan. If he does, Novak Djokovic may have good cause to complain.

A Grand Slam tennis draw consists of 128 players, arranged in a knockout tournament. Of course the draw is not totally random. The top 32 players are seeded and the draw is arranged so that these seeded players can only meet in the later rounds.

It is questionable why Djokovic is the top seed at the French Open, given that Rafael Nadal has the clear edge on the clay courts of Roland Garros. Seeding players, however, is complicated and inherently subjective, making it difficult to avoid such quirks. Still, there are some very strange things afoot.

Being the top two seeds, Djokovic and Nadal are placed in separate halves of the draw. Similarly, the top four seeds are in their own quarters, the top eight seeds are in their separate eighths, and so on. Finally, with the 32 seeds in place, the remaining 96 spots are to be randomly allotted to the cannon fodder (with all due respect to Gustavo Kuerten). In reality, it appears that the assigning of those final spots is not always as random as is claimed, which makes for an interesting story. We'll focus upon the top players, however.

Barring upsets, the seeded players will eventually meet, but what then? Suppose, for example, that the top four players make it through to the semifinals. Who, then, is to play whom? That's when things get weird.

There are three ways that the top four seeds of a tournament might be scheduled to meet, of which the most natural pairing would appear to be 1 v 4 and 2 v 3. A second possibility, 1 v 2 and 3 v 4, is wisely precluded by the tennis rules.

What about the third possibility, with 1 v 3 and 2 v 4? It's difficult to see any argument for this arrangement, which typically advantages the second seed over the first. Nonetheless, for a few years, long ago, Aussie rules football used something like this seeding for its finals.[1] Grand Slam tennis tournaments employ a system almost as peculiar.

According to the Grand Slam rules, the 3rd and 4th seeds are placed in distinct quarters of the draw and away from the two top seeds, but otherwise are placed randomly. (There is then a similar random process used to place the lower seeds.) This means that there is a 50-50 chance that the draw will point towards semifinals of 1 v 3 and 2 v 4.

Such dice-rolling seem remarkably amateurish for a multi-zillion dollar sport. And, as the dice would have it, in the current French Open the top seed Djokovic is scheduled to meet 3rd seeded Federer, while Nadal can look ahead to the significantly easier (if not easy) 4th seeded Andy Murray. Perhaps that is justice, since arguably Rafa should be top seed. Nonetheless, Novak appears to have good reason to feel aggrieved.

But maybe the Grand Slam bosses know more than it seems? In what sense can we be sure that 1 v 4 and 2 v 3 is the best pairing? The question is surprisingly tricky.[2]

Let's assume that the seeding is accurate, that Djokovic is stronger than Nadal, who is stronger than Federer, who is stronger than Murray. We'll also assume that every player has less chance of beating a higher seed than a lower seed; for example, Murray has less chance of beating Djokovic than he has of beating Nadal. With these assumptions, 1 v 4 and 2 v 3 can be shown to be a natural pairing in a number of respects. In particular, Djokovic has a greater overall chance of winning than Nadal, who has the next greatest chance, and so on. Moreover, 1 v 4 and 2 v 3 is the only pairing that is always fair in this sense.

Surprisingly, however, this natural pairing does *not* necessarily give the greatest chance of a Djokovic-Nadal final. Suppose that the winning chances between the four players are as follows:

	D	N	F	M
D		4/7	5/7	6/7
N	3/7		4/7	5/7
F	2/7	3/7		4/7
M	1/7	2/7	3/7	

The table indicates, for example, that Nadal has a 4/7 chance of beating Federer, and conversely Federer has a 3/7 chance of beating Nadal. In this scenario, the pairing 1 v 4 and 2 v 3 gives a 24/49 chance of a Djokovic-Nadal final. The

[1]The "rules" in Aussie rules should be thought of in the Calvinball sense, and on occasion the finals systems were designed in the same spirit.

[2]For an extended discussion, see T. Vu and Y. Shoam, *Fair Seeding in Knockout Tournaments*, ACM Transactions on Intelligent Systems and Technology, **3**, 2011.

pairing 1 v 3 and 2 v 4, however, does a tiny bit better, giving a 25/49 chance of the dream final.

Analyzing the top eight seeds requires considerably more work: there are 315 essentially different arrangements of the eight seeds in the quarter finals, of which 48 are consistent with the Grand Slam rules.[3] And, many strange things can happen with these arrangements.

The most natural pairing is usually considered to be 1 v 8, 4 v 5, 2 v 7 and 3 v 6, but it is difficult to say why. Many things can go wrong, and in particular this pairing is not always fair in the sense described above.

	D	N	F	M	5	6	7	8
D		.55	.55	.55	.55	.95	.95	.95
N	.45		.55	.55	.55	.95	.95	.95
F	.45	.45		.55	.55	.55	.95	.95
M	.45	.45	.45		.55	.55	.95	.95
5	.45	.45	.45	.45		.55	.95	.95
6	.05	.05	.45	.45	.45		.95	.95
7	.05	.05	.05	.05	.05	.05		.95
8	.05	.05	.05	.05	.05	.05	.05	

Imagine the winning chances of the top eight seeds against each other are as in the table above. Then the "natural pairing" will actually give second seed Nadal about a 35% chance of winning the final, with top seed Djokovic only a 30% chance.

So, maybe the Grand Slam chiefs are cleverer than we thought. Perhaps all that randomness in the draws is a brilliant way to cope with some tricky mathematics. Perhaps ... Nah. It's still a silly idea and Djokovic still has every right to be grumpy.

Puzzle to ponder

Show that there are 315 essentially different ways to arrange the top eight seeds of a tournament, and show that 48 of these ways comply with the Grand Slam rules.

[3]The "essentially" refers to the fact that what really matters is not the seeding but the pairings, who plays whom. For example, in a tournament where 1 v 4 and 2 v 3, it doesn't matter whether Djokovic is seeded 1 and Murray 4, or vice versa.

CHAPTER 24

Tennis math, anyone?

Recently, we wrote about seeding at the French Open, very concerned that top seed Novak Djokovic was being treated unfairly.[1] In the end it didn't matter: predictably, Djokovic and second seed Rafael Nadal won through to the final and, very predictably, Nadal was the victor. That was despite the umpire unsettling Rafa, by demanding that the finalists play on through hell and absurdly high water.

Now it's on to Wimbledon and more wonderful weather, although at least the English have sufficient sense to come in out of the rain. This means that Djocky and Rafa will probably have lots of time to while away in the locker room. So, how might they occupy themselves? Well, math is always a good option.

Here's an easy question to start off our tennis champs: in total, how many matches will be played in the Men's Singles at Wimbledon?

There are 128 players in the draw, which means there are 64 matches in the first round. The 64 winners then meet in the 32 matches of the second round, and so on until, inevitably, Djokovic and Nadal are left to play the final match.[2] So, we just have to sum $64 + 32 + 16 + 8 + 4 + 2 + 1$.

Simple stuff. However, there is an even simpler method to obtain the answer. There are 128 players, and all but one of them – our money's on Djokovic – will be eliminated along the way, one match at a time. So, there must be 127 matches in total. Very, very easy.

The original sum, with each new number a fixed multiple of the previous number, is what is known as a *geometric series*. In our case the fixed multiple is 1/2. And, Wimbledon has provided us with an easy method to find the sum of our particular geometric series:

$$64 + 32 + 16 + 8 + 4 + 2 + 1 = 128 - 1\,.$$

Can we similarly do other sums? Wimbledon begins with 128 players, which is 2^7, and definitely other powers of 2 are no problem. If we begin with 2^N players, who compete in pairs in the usual manner, then counting the number of matches in the two different ways shows that

$$2^{N-1} + 2^{N-2} + \cdots + 4 + 2 + 1 = 2^N - 1\,.2$$

[1]See the previous Chapter.

[2]It turned out that Roger and Andy had other ideas.

What about other powers? In a short and lovely article, mathematician Vincent Schielack used the tournament scenario to derive formulas for other geometric sums.[3]

To apply Vincent's very clever argument, let's imagine we have a new television show, *Voicing with the Stars.* The show begins with 2187 contestants, who are separated into 729 groups of three. Each group has a sing-off to see who advances. In the next round (yawn!), 729 contestants are left, who are again grouped into threes for an elimination sing-off. Eventually, just one singer – our new Kylie Minogue – remains. So, how many sing-offs would we have had along the way?

As with Wimbledon, there are two different ways to count the number of "matches". Summing directly, we have 729 sing-offs in the first round, then 243 sing-offs in the second round, and so on, dividing by 3 each time. However, overall we know that all but one of the 2187 contestants will be eliminated, with two contestants eliminated in each sing-off. Equating these two calculations of the number of sing-offs, we find that

$$729 + 243 + 81 + 27 + 9 + 3 + 1 = \frac{2187 - 1}{2}\,.$$

An analogous formula then holds if we begin with 3^N singers. And it's just as easy to begin with M^N singers, with sing-offs between groups of M singers, and with one member from each group progressing to the next round. Arguing just as we have above, you should be able to convince yourself that

$$M^{N-1} + M^{N-2} + \cdots + M^2 + M + 1 = \frac{M^N - 1}{M - 1}\,.$$

So for any positive whole number M, other than 1, we have a formula for the geometric sum with constant multiple $1/M$. Or, reading the sum from right to left, we can think of the constant multiple as the whole number M. But what about other values of M? Is the formula always true?

If $M = 1$ then the formula has a troublesome 0 in the denominator, but in this case we hardly need a formula: the sum reduces to adding 1 to itself a bunch of times. For any other value of M the formula is indeed true. How can we see that?

If M is a fraction then we can still argue with tournaments or TV shows to prove the formula. Alternatively, there is a well-known algebraic proof that will justify the formula for other M.[4] However, as Victor points out, we don't really need to argue further. Given our geometric sum formula is true for the infinitely many integers beyond 1, the formula will also automatically be true for all numbers other than 1. That's not overly difficult to justify, but we don't want to give the whole game away: let's leave something for Djocky and Rafa to ponder.[5]

Puzzle to ponder

Can you find a "tournament" scenario that can show that our geometric formula also holds for negative integers, $M = -1, -2, \cdots$?

[3] *Summing Geometric Series by Holding a Tournament*, College Mathematics Journal, **23**, 210–211, 1992.

[4] This is the standard proof that would normally be given in school, and is easily found online.

[5] The point is that both sides of our geometric sum for formula are "nice", except at $M = 1$. The technical term is *analytic*, and there is a general theorem that if two analytic functions are equal on infinitely many points then they are equal everywhere (except perhaps explosion points, like $M = 1$).

CHAPTER 25

The ball was in AND out? You cannot be serious!

Wimbledon begins tonight, making your Maths Masters nostalgic for John McEnroe. We remember McEnroe's epic battles with Björn Borg, and also his terrific tantrums. But just imagine McEnroe's reaction if he had been told that Borg's serve was definitely out, but that no linesman was permitted to call it. Can we be serious?

The following diagram shows the skidmarks of some of Borg's serves from the far side into the near right-hand court. The brown marks indicate serves that were in and the orange marks indicate faults. We've also identified the sharp-eyed linesmen keeping watch on the three lines bounding the service box.

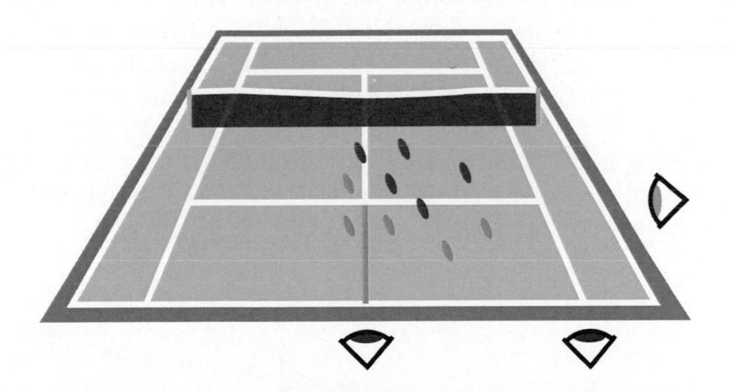

Each linesman checks the ball only in relation to his line. He yells "FAULT!" if the ball is completely on the wrong side of his line and remains silent otherwise. If all three linesmen are silent then the ball is taken to be in.

Now consider the diagram below, showing the yellow skidmark of one of Borg's faults. The purple linesmen will definitely keep silent since the ball is far to the correct side of his line. But both the red and blue linesmen will view part of the ball as level with their line, and so they will keep silent as well. So, though the ball is definitely a fault, no linesman is permitted to call it. At which point, McEnroe presumably hurls his racquet.

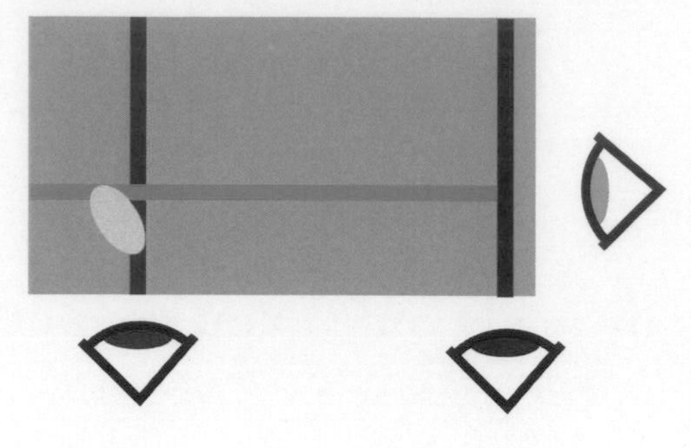

Of course, balls are much more commonly miscalled simply as a result of human error. Which is why the infallible computer system known as Hawk-Eye was introduced. How do we know Hawk-Eye is infallible? Because the tennis commentators constantly act as if Hawk-Eye is infallible. Hmm.

Hawk-Eye will occasionally get it wrong, of course, notwithstanding some confusing hype from Hawk-Eye's inventors. Hawk-Eye takes video footage from up to ten high-speed cameras and uses the footage to reconstruct the trajectory of the ball. But any such filming and reconstruction will always contain some approximation and some error.

The International Tennis Federation demands that any electronic system must never be in error by more than 10 millimeters, and that the average error must be below 5 millimeters. It is claimed that Hawk-eye is accurate well within these tolerances. As British academics Harry Collins and Robert Evans have noted, however, precise details of the testing of Hawk-Eye are not made available.[1] This makes it difficult to know exactly what is true, or even what "average error" means in the context of these tests.

We have little doubt that, though not perfect, Hawk-Eye is in general very accurate. However, we do not believe that Hawk-Eye will always be accurate to within 10 millimeters. With just the wrong ball trajectory, Hawk-Eye could definitely be out by centimeters. Imagine a ball shooting across the ground at a very shallow trajectory. Here, even tiny errors in Hawk-Eye's calculations can create huge errors in determining the precise location of where the ball actually touches the ground. And of course Wimbledon's grass court, with its natural unevenness, does not make things any easier.

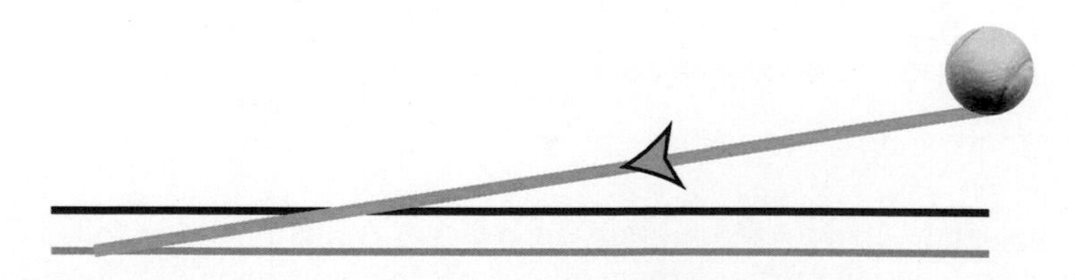

You might argue that the high tennis net rules out such skimming shots. However, since balls can also be hit around the net, not just over it, these troublesome trajectories can definitely occur. In such situations it is not clear that Hawk-Eye will be more accurate than the linesmen.

[1] *You Cannot Be Serious! Public Understanding of Technology With Special Reference to "Hawk-Eye"*, Public Understanding of Science, **17**, 283–308, 2008.

Obviously such occurrences are rare, and Hawk-Eye is in general of great assistance to quality linecalling. But we do get annoyed when gullible commentators treat video simulations as if they were reality. We wish we had McEnroe by our side to say it for us: they cannot be serious!

Puzzle to ponder

How might you rule on these tricky balls to avoid the paradoxical situation considered here?

CHAPTER 26

Giving it your best shot

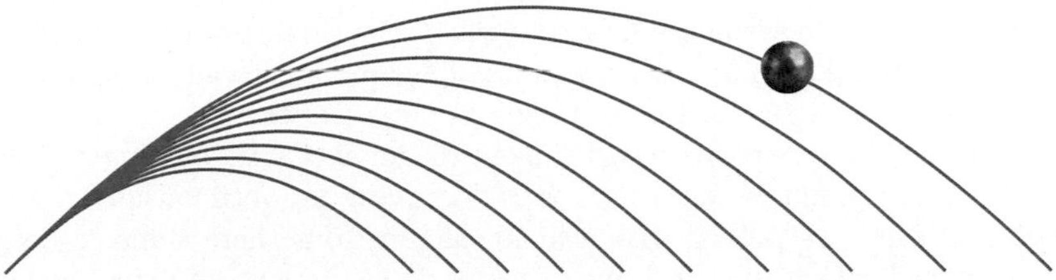

How do you become a great shot putter? Work hard and eat your Wheaties. Then you might follow in the footsteps of Justin Anlezark, representing Australia in three Olympic Games. And, maybe applying a little mathematics will help get you there.

Shot putting is a great application of the mathematics of projectiles. We launch the iron shot with a vertical speed V and a horizontal speed H. Then the Earth's gravitational acceleration g determines when and, most importantly for a shot putter, where the shot will land.

If we launch from the ground with vertical speed V then the shot will hit the ground with that same speed V downwards. From that, the acceleration determines the time in flight to be $2V/g$. Multiplying this time by the horizontal speed H, we find that the shot lands away a distance $2HV/g$. These calculations pretend there's no air resistance, but the high density of the 16 pound shot means that this is a reasonable assumption.

How do we make the distance traveled as large as possible? Obviously, by throwing harder (hence the Wheaties). But given we can only throw at some maximum total speed, our task is to find the optimal launch angle. Most of us have experimented with this: just recall trying to douse your annoying brother with the garden hose, and struggling to get the water stream to reach as far as possible.

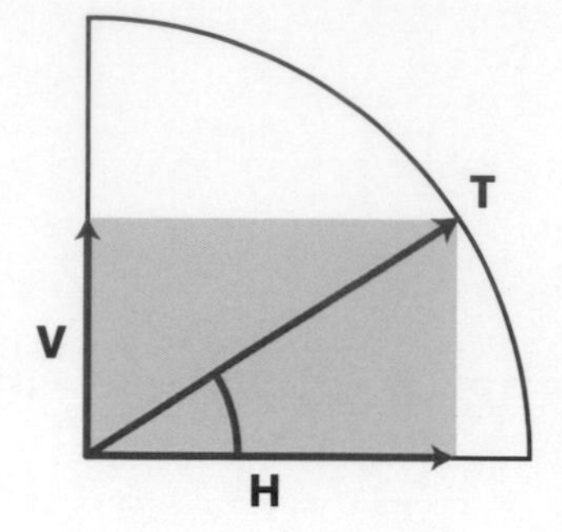

Forming a rectangle with base H and height V, the total speed T is the diagonal. So, making the distance $2HV/g$ as large as possible amounts to adjusting the launch angle to make the area HV of this rectangle as large as possible.

You may not be surprised that, for fixed diagonal T, the rectangle of largest area is the square. That is, we ensure $H = V$, giving an optimal launch angle of 45°. This can easily be proved with a little calculus, but there is also a beautiful elementary proof, which we will leave as a puzzle (Note also, if your brother has commandeered the hose then your strategy is to convince him that the best angle is 90°).

These calculations are not completely realistic. First of all, an actual shot putter will launch the shot about 2 meters above the ground. The consequence is that it is better to launch at a slightly shallower angle. At the speed shot putters throw, the optimal launch angle is about 42°.

Secondly, shot put is not just mathematics. There are also biomechanical considerations. It turns out that shot putters can throw harder at shallower angles. So, there is a trade-off, with most shot putters throwing around 35°.

In fact, Justin Anlezark is unusual in that he throws close to the mathematical optimum angle of 42°. And he throws well enough to be a serious medal chance in Beijing. So we asked Justin, does he use mathematics to analyze his technique? "Nah. If you throw 20 meters at 42°, or you throw 20 meters at 30°, you just think 'Hey, I threw 20 meters!' "

Puzzle to ponder

Prove that, for launching the shot from the ground, the optimal angle is 45°.

CHAPTER 27

Bombs, and a bombed Riewoldt

The national curriculum is of course very important, and it has occupied us for quite some time.[1] There is now a much more compelling issue, however: Nick Riewoldt's hamstring.[2]

It's not just Saint Nick. In the past two weeks there have been seven hamstring injuries in the AFL,[3] three of them more or less season-ending. People are asking why, with many fingers pointed at the speed of modern football. There have been loud calls to limit the number of substitutions in AFL games.

Is this reaction warranted? Will we hear more hamstrings twang in today's big game? Or, has this all been just a streak of bad luck?

Statisticians (which definitely does not include us) love to tackle such questions, and we'll return to the hamstrings later. First we want to consider a different scenario, more in keeping with the weekend's Anzac remembrance,[4] and even more important than Riewoldt's hamstring: the World War II bombing of London.

There was no shortage of attacks on London from conventional German bombers, especially early in the war. From 1944, however, the majority of attacks were in the form of unmanned V-1 and V-2 flying bombs. Throughout the war, London was struck by thousands of such bombs, and some parts of the city

[1] Australia has had its own version of the Math Wars, although it has been more of a capitulation than a war. See Part 12 of *A Dingo Ate My Math Book*.

[2] At the time, Nick Riewoldt was the star player for the Aussie rules football team, St. Kilda.

[3] The Australian Football League, the major league for Aussie rules.

[4] Anzac Day, celebrated each April 25, is essentially Australia's version of America's Memorial Day, to remember those who served and died in wars.

were struck much more frequently than others. This suggested that the Germans were choosing targets and were able to accurately aim their bombs to hit them. Of course a prudent Londoner would look to avoid these targets.

British statisticians wanted to determine if the Germans were in fact able to aim their bombs that precisely, or if the bombs were falling more randomly. To do so, they employed a simple test. To set it up, they divided a region of London into a 24×24 grid, consisting of 576 squares. The statisticians then counted the number of bombs that hit each square over a certain period.

During that period the total region was hit by 537 bombs, meaning each square in the grid was hit an average of $537/576 = 0.932$ times. That's enough bombs for most squares to be hit, and if the bombs were falling randomly then perhaps one might expect that. In fact, 229 of the 537 squares were not hit at all, which is intuitively not very random. But, intuition can be deceiving.

If the bombs were actually falling randomly then the clustering of the strikes would follow what is known as a *Poisson distribution.* This distribution predicts that the number of squares hit by exactly N bombs should be approximately

$$576\, e^{-0.932} \frac{0.932^N}{N!}.$$

In this formula, $e = 2.718281828\cdots$ is that other famous irrational number, the "natural" base for logarithms, and $N!$ is the *factorial* product $1 \times 2 \times 3 \times \cdots . \times N$. (0 is a special case, and it turns out to be convenient to define $0! = 1$).

For example, setting $N = 0$ the formula predicts that about 227 squares will fail to be hit. That is remarkably close to the actual number of 229.

number of hits	**predicted number of squares**	**actual number of squares**
0	227	229
1	211	211
2	99	93
3	31	35
4	7	7
5+	2	1

The complete table is given above, with the Poisson predictions being strikingly close to the actual numbers. This convinced the statisticians that the bombs were indeed falling randomly, and so there was no point in attempting to shuffle Londoners into "safe" zones. Given the deceiving nature of intuition, we wonder if the bomb-dodging Londoners were so readily convinced.

Now, back to our footballers and their hamstrings. Hamstring injuries are a huge issue for the AFL, responsible for many more lost games than any other injury type. Specialists in sports medicine work tirelessly to improve prevention and treatment. So, if the frequency or severity of such injuries is changing, that is big news. How can we tell if that is the case? How good or bad is 7 hamstring injuries in the previous fortnight? What about the (roughly) 13 hamstring injuries total in the four weeks to date?

A hamstring can go twang at any time. They may not be random like the German bombs, but it is reasonable to begin with that assumption. That suggests that the Poisson distribution might provide us some insight.

The AFL publishes very detailed summaries of player injuries. The most recent is the 2008 injury survey, which indicates that there were 106 hamstring injuries over the 22 week season. That tells us that on average there were about 9.64 hamstring injuries each fortnight.

Clearly, just on average, the 7 injuries of the last fortnight is low. In fact, based on the 2008 frequency, the Poisson formula (with no 576 there, and with 9.64 taking over the role of 0.932) suggests that there's about a 25% chance of 7 or fewer hamstring injuries in a given fortnight. Moreover, Poisson (now with $2 \times 9.64 = 19.28$ in the starring role) indicates there is about a 9% chance of having 13 or fewer injuries in a given four-week period.

For us, this is tentative evidence that the frequency of hamstring injuries in 2010 is genuinely lower than the 2008 frequency. Of course, that is cold comfort to Saints fans, staring at a season with Riewoldt on the sidelines.

What we're obviously leaving out is any analysis of the severity of the hamstrings. That would be more involved, and the AFL survey doesn't contain the data required. We're honestly not sure where such an analysis might lead.

In any case, for Saints fans the analysis is very easy: one hamstring, no superstar, and the probability of a premiership having plummeted.

Puzzle to ponder

Suppose 4 bombs are randomly dropped on a grid containing just 4 squares. What is the *exact* probability that a given square will not be hit? How does that compare to the Poisson distribution estimate of this probability? What if we had 16 bombs dropping on a 16-square grid?

CHAPTER 28

Diophantine footy fan

We've already had some great AFL games this season. It kicked off with the Giants' huge upset of the Swans, the Dons and the Pies fought their traditional Anzac Day battle and Port Adelaide pummelled the Cats.[1] The highlight of the year, however, was in Round 4, when Hawthorn met the Gold Coast Suns.

Say what? Sure, watching a 99 point thrashing would have been great fun for Hawks supporters but why should anyone else get excited?

The nerdier footy fans would have noted that the Suns finished with a score of 7.7, that is, 7 goals and 7 behinds. For anyone just off the boat,[2] a goal in Aussie rules is worth 6 points and a behind is worth 1 point. That means the Suns' total score was $(7 \times 6) + 7$, amounting to 49. But 49 is also 7×7, the product of the Suns' goals and behinds. Amazing.

Well, not exactly amazing, but it is cute. It's a favorite score amongst mathematically inclined footy fans.

Unsurprisingly, this was not the first time that a "total equals product" score had been kicked. Back in 1897, in the very first round of the VFL,[3] South Melbourne kicked 3.9, for a total of 27. Such scores are not all that common, however. None occurred in 2013, much to the chagrin of your Maths Masters, who had patiently waited all season for the trigger to write this column. It occurred only once in 2012, the Suns again kicking 7.7.

---

[1] Anzac Day is essentially Australia's version of America's Memorial Day. It's now also a big day for sports.

[2] Or, for any non-Australians ...

[3] The Victorian Football League, the long-running precursor to the current Australian Football League

Let's look further. We know 7.7 equals 49 and 3.9 equals 27, but are there any other scores that will work? Supposing a team kicks G goals and B behinds, their total score is then $6G + B$. We're then asking if that total equals the product $G \times B$. So, we're looking for solutions to the equation

$$6G + B = G \times B\,.$$

That looks like a very easy problem. We can seemingly just choose one of the numbers to be pretty much anything, and then solve for the other.

Let's try it with $B = 10$, which means we then have to solve $6G + 10 = 10G$. That gives us $4G = 10$, and so $G = 5/2$. Hmm. On a bad day the Gold Coast might wind up with 2 1/2 goals, but it generally seems unlikely.

Of course we are only interested in nonnegative whole number solutions, which means we need to be a little cleverer. Subtracting B from both sides of the above equation, our problem is then to solve

$$6G = (G - 1)B\,.$$

This new equation may not appear any simpler but having both sides factored turns out to be the key. We now know that $G - 1$ must divide evenly into $6G$. However it is impossible for G and $G - 1$ to have a common factor other than 1, which means that $G - 1$ must divide evenly into 6. The factors of 6 are 1, 2, 3 and 6, each of which leads to one of our special scores:

G	B	Total
2	12	24
3	9	27
4	8	32
7	7	49

Actually, there is one more solution to our footy puzzle: if G and B are both zero, leaving the team scoreless. (It's worthwhile pondering how this solution escaped our calculations above.) That has never occurred in the AFL/VFL, although in 1899 St. Kilda gave it the old college try, scoring just one point in a game against Geelong.

The above puzzle was fun but was there any real purpose? Well, if you're a poor St. Kilda fan, yet again watching the not-so-mighty Saints being thrashed, solving such puzzles can while away the depressing hours. But there is no deeper meaning.

Although our footy puzzle is just a bit of fun, however, it is nonetheless an example of a very important type of problem, known as a *Diophantine problem.* These problems take their name from the Greek mathematician Diophantus, who is celebrating his 1800th birthday this year (give or take a couple hundred years). He is considered to be the father of algebra, being the first known mathematician to use letters to stand for unknown numbers, just as we have above.

A Diophantine problem involves determining all the whole number solutions to an equation, or to a collection of equations. The equation itself may be quite simple but Diophantine problems can be incredibly difficult. Testament to this is *Fermat's last theorem*, the most infamous of Diophantine problems, and taking 350 years to solve.[4] Even Pythagoras's theorem leads to a tricky Diophantine problem: there are zillions of right-angled triangles, but how do we find all the triangles with whole number sides?

One of the oldest, most famous and most astonishing of Diophantine problems is Archimedes' cattle problem. (The great Archimedes lived around 250 BC; he wasn't the father of algebra but he was the father of pretty much everything else in mathematics.) You have cows and bulls of four different colors, and the problem is to determine the number of cattle of each type. Extracting the math from Archimedes' words, the relationships between the types of animals turns out to be given by the following nine equations:

$$\begin{aligned}
\text{white bulls} &= (1/2 + 1/3)(\text{black bulls}) + (\text{yellow bulls})\\
\text{black bulls} &= (1/4 + 1/5)(\text{dappled bulls}) + (\text{yellow bulls})\\
\text{dappled bulls} &= (1/6 + 1/7)(\text{white bulls}) + (\text{yellow bulls})\\
\text{white cows} &= (1/3 + 1/4)(\text{black herd})\\
\text{black cows} &= (1/4 + 1/5)(\text{dappled herd})\\
\text{dappled cows} &= (1/5 + 1/6)(\text{yellow herd})\\
\text{yellow cows} &= (1/6 + 1/7)(\text{white herd})\\
\text{white bulls} + \text{black bulls} &= \text{a square number}\\
\text{dappled bulls} + \text{yellow bulls} &= \text{a triangular number}
\end{aligned}$$

The reader is invited to hunt for a solution to the problem. But please note: if we omit the last two equations the smallest solution to the resulting much easier problem requires over 50 million animals. The complete problem results in an absolutely astronomical number of animals.

It may not be surprising that Archimedes' cattle number is in the millions but even a single, simple Diophantine equation may only have huge solutions. The equation $x^2 - 61y^2 = 1$ was considered by the 12th century Indian mathematician Bhaskara. (Equations of this form are known as *Pell equations*; they are very important and play a critical role in solving the complete cattle problem.)

[4]Fermat's last theorem states that the equation $x^n + y^n = z^n$ has no solutions for x, y and z positive integers, and n a positive integer greater than 2. Famously, the French mathematician Pierre de Fermat claimed in the margin of a notebook to have a proof, but no such proof was ever found in his writings. The theorem was finally proved in 1994, by Andrew Wiles.

Bhaskara found the smallest solution of this equation, with $y = 226,153,980$ and $x = 1,766,319,049$.

So, perhaps we're lucky that our footy equation turns out to have only small solutions. If the totals were in the thousands, or higher, there would no hope that a team would ever actually achieve such a score. Unless, perhaps, they were playing St. Kilda.

Puzzles to ponder

Show that if G is a positive whole number then G and $G-1$ have no common factor other than 1.

Suppose a goal is worth 5 points, rather than 6. Find all "total equals the product" scores with this new scoring system.

CHAPTER 29

Walk, don't run!

We all know how to walk. So, it shouldn't be too difficult to state precisely what walking is. Then we can have a race and see what happens.

That of course has been done, and the results have been entertaining, controversial and confusing. Race walking began over 300 years ago as *pedestrianism*, the charming practice of English lords pitting their footmen against each other. As the sport gained in popularity, and voluntariness, efforts were made to clarify exactly what constitutes walking, to preclude cheating by what amounts to running. And thus began a long history of everchanging rules.

The rules for athletics events are now determined by the International Association of Athletics Federations. Its current definition of race walking contains two understandably pedantic parts:

1. *Race Walking is a progression of steps so taken that the walker makes contact with the ground, so that no visible (to the human eye) loss of contact occurs;*

2. *The advancing leg shall be straightened (i.e. not bent at the knee) from the moment of first contact with the ground until the vertical upright position.*

Clumsy verbiage aside, the IAAF definition pretty much describes how your legs function during a leisurely walk. Then, if you speed up, you'll start bending your knees and your feet will begin to lose contact with the ground. That is, you'll be running.

The first IAAF rule will definitely stop most contestants from running, and the second rule is intended to prohibit weird shuffling that is difficult to distinguish from running. But would you actually refer to the motion of race walkers as walking?

At the top of our column is a photo of Australia's Jared Tallent in full waddling stride. Jared won bronze and silver medals at the Beijing Olympic Games, and is competing again in the London Games. So, there must be some benefit to Jared's strange gait.

Let's first imagine Jared walking in a more familiar, waddle-free manner. According to the second IAAF rule, Jared's weight-bearing leg must be rigid from the moment his foot touches the ground until his leg is vertical. The result of this is that Jared's "centre of mass" (located slightly above the top of his legs) will be rotating around his anchored foot.

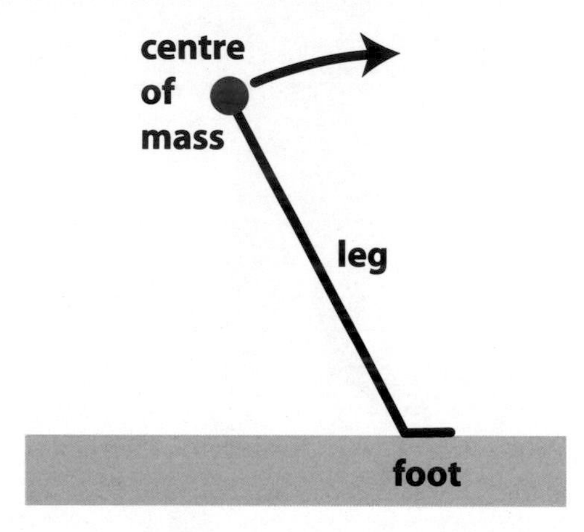

However, objects don't just rotate for no reason. The (roughly) circular motion of the planets, for instance, is a consequence of the gravitational force of the Sun. And similarly, the rotation of Jared's body around his anchored foot is due to Jared's weight, the Earth's gravitational force pressing Jarod's body down on his leg.

It follows that if Jared is moving too quickly then there will be insufficient gravitational force for this rotation, and Jared's foot will lift from the ground. Using familiar formulas for circular motion we can estimate this lift-off speed to be about 12 kilometers per hour.

We can now compare this "maximum" speed of 12 kilometers per hour to the actual speeds of race walkers. For example, the world record for the 20 kilometer walk is 1 hour, 17 minutes and 16 seconds, held by Russian Vladimir Kanaykin. That is an average speed of over 15 kilometers per hour.

How did Vladimir manage such a speed? Well, EPO may have helped: six months after creating his world record Vladimir was banned for a positive drugs test. However, though the EPO may help explain Vladimir's world record, it cannot explain his maximum speed: taking proscribed drugs is one way to violate the laws of athletics but it cannot help you violate the laws of physics.

This is where the waddle comes in. As Jared's leg rotates and his hip rises, Jared twists to allow his opposite hip to drop down. The underlying purpose is to lower Jared's centre of mass, to a roughly constant height, meaning less force is required for the rotation. The effect is clearly visible in videos of race walkers, demonstrating that their bodies do not bob up and down in the manner of runners' bodies.

But sadly, the waddle is not the solution to all of race walkers' worries. One of the most heartbreaking moments of the 2000 Sydney Olympics was Australian Jane Saville being red carded for running, just meters before entering the stadium for her victory lap. She was disqualified for having been seen to "fly" on three different occasions by three different judges.

Was Jane trying to cheat? Almost certainly not. Notice that it is permitted for the "walker" to lose contact with the ground, just as long as the loss of contact is not visible to the (human) judge. And, detailed investigations have shown that lifting happens continually in races, and is effectively unavoidable at the very high speeds of international competitions. It is just that most of this lifting is so brief as to go undetected: a judge can typically only detect a lifting violation if it exceeds 1/25 of a second.

This leaves us with a very unsatisfactory state of affairs. Since lifting occurs on a regular basis, we're probably no longer dealing with walking. Furthermore, the unpredictable manner in which judges detect lifting makes any race very much a lottery. The resolution probably lies in a technological fix, for instance by attaching sensors to the walker's shoes. However, any such imposition of technology will have dramatic repercussions on walkers' styles.

In any case, we'll be cheering for Jared Tallent, and Australia's other representatives in the Olympics walking events. We wish them best of luck with the judging lottery, and of course with their walking, or running, or waddling, or whatever exactly it is that they do.[1]

Puzzle to ponder

For what distance can Jared move with both feet off the ground, without it being detected by a judge?

[1] Jarred scored a silver medal in the 50 km walk at the 2012 London Olympics. This was then upgraded to gold in 2016 when Russian Sergey Kirdyapkin was stripped of his medals for doping.

CHAPTER 30

Tour de math

From an online cycling column: *"If you want even the nerds to consider you a nerd, try getting enthusiastic about bicycle gearing."* Which presumably makes us the nerds' nerds. It's Tour de France time, and we take it to be the ideal time to discuss gear ratios.

What's a gear ratio? Consider, for example, the sprocket and chain system pictured above. The large sprocket has 22 teeth and the small one has 7 teeth. Then the gear ratio is simply the ratio of teeth, in this case cleverly arranged to be 22/7, or about 3.14. Imagine that, as is usual, the large sprocket is attached to the bike pedals, and the small sprocket is attached to the back wheel. Then pedaling one revolution results in 3.14 revolutions of the back wheel.

Most commercial bikes have a number of both large and small sprockets, giving a seemingly large range of gear ratios. For example, our Maths Master CyclePro has sprockets with 28, 38, and 48 teeth attached to the pedals, and sprockets with 14, 16, 18, 20, 22, 24, and 28 teeth attached to the back wheel. So, our bike has $3 \times 7 = 21$ different gears? Well, no, it hasn't.

Let's have a look at the table of the 21 different possible gear ratios.

	14	**16**	**18**	**20**	**22**	**24**	**28**
28	**2.00**	**1.75**	**1.56**	**1.40**	**1.27**	**1.17**	**1.00**
38	**2.71**	**2.38**	**2.11**	**1.90**	**1.73**	**1.58**	**1.36**
48	**3.43**	**3.00**	**2.67**	**2.40**	**2.18**	**2.00**	**1.71**

As you can see, the gear ratio of 2 appears twice. What this means is that riding the bike blindfolded (do not try this at home), you could not distinguish the gears 48-24 and 28-14. As well, a number of the other gear ratios are very close to being repeated. For all practical purposes, our 21-gear bike has only 14 distinguishable gears.

Does this mean that the Maths Masters have been ripped off by some conniving bike dealer? If so, we are in good company. Below is the gear ratio table for one of the bikes of Tour de France legend, Lance Armstrong.[1]

	11	12	13	14	15	16	17	18	19	21
39	3.55	3.25	3.00	2.79	2.60	2.44	2.29	2.17	2.05	1.86
53	4.82	4.42	4.08	3.79	3.53	3.31	3.12	2.94	2.79	2.52

It turns out that duplications in gear ratios are pretty much unavoidable. To begin, mechanical considerations rule out some sprocket combinations as impracticable: using a bad combination of sprockets results in twisting of the chain, causing excessive wear.

For someone in the heat of a race, there is also a more immediate concern. It is usually infeasible to change gears in the order suggested by their gear ratios; doing so would involve a complicated shifting of both front and rear sprockets. Instead, most practical shifting of gears involves adjusting the rear sprocket only, with the front sprocket adjusted only occasionally.

There is one last gear problem we'd love to solve. Being bike nerds, we would love to replace our original gear ratio of $22/7$ by a gear ratio of exactly π. Alas, we are doomed to fail. That π is an irrational number tells us exactly that this precise gear ratio is impossible.

But not all hope is lost. Returning to a bygone era, we have been tinkering with the relative wheel sizes on penny farthings, getting the proportions just right. The result is our nerdishly mathematical masterpiece. Ladies and Gentleman, we proudly present to you: the picycle.

Puzzle to ponder

Design a system of gears that gives the ratios 1.2, 1.5, 1.6, 2, 2.4, 2.5, and 3, and only these ratios.

[1] Of course, Lance is now legendary for much less heroic reasons.

CHAPTER 31

How round is your soccer ball?

It's time for the World Cup! So we're going to need plenty of good, round soccer balls. And how to make a perfectly round sphere? Blow a soap bubble, of course.

Unfortunately, Harry Kewell has steadfastly refused to use soap bubbles for his penalty kicks.[1] An alternative is to work with the Famous Five regular solids. First, sew together some flat leather polygons to form the solids. Then, inflating the polyhedrons, the sides of the solids will bulge into a roughly spherical form.

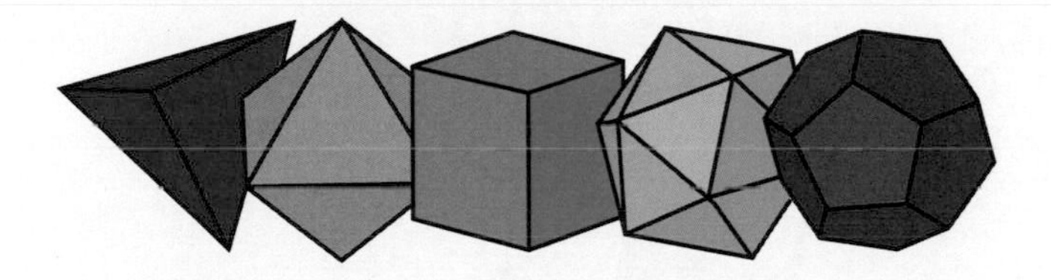

Predictably, this works poorly for the cube and the two pointier solids. The blue icosahedron and the purple dodecahedron, however, work reasonably well. In fact, we own a commercially made dodecahedral ball.

A standard soccer ball is a variation of this idea. It consists of twelve pentagonal and twenty hexagonal panels. It can be thought of as an icosahedron with its twelve corners sliced off, and with patches sewn over the resulting pentagonal holes.

[1] Harry Kewell was Australia's star winger at the time.

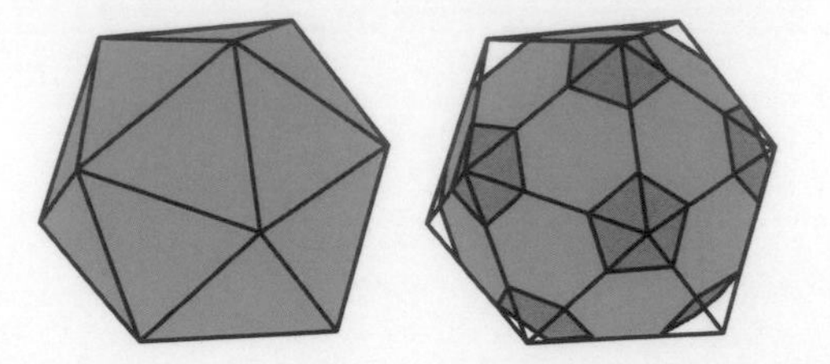

There is still a decision to be made: how large should you make the pentagonal holes and patches? A natural mathematical approach is to ensure all the edges of all the panels have the same length. However, this makes the hexagons significantly larger in area than the pentagons, which in turn results in markedly different bulging of the two types of panels.

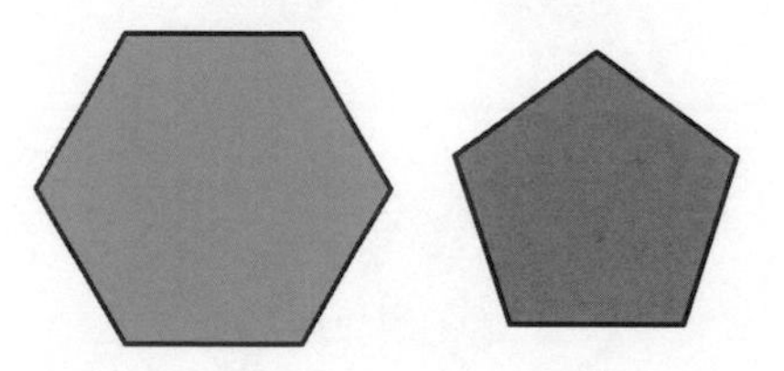

To even out the bulging, better quality balls are constructed using slightly larger pentagons. That requires the sides of the hexagons to have two slightly different lengths, and so the hexagons are no longer perfectly symmetric.

After all that, you finally have your soccer ball. Will FIFA, the governing body for soccer, judge your ball sufficiently round for a good game of soccer? A soccer ball consists of $20+12 = 32$ panels, each panel facing a similar panel on the opposite side of the ball. That gives sixteen pairs of panels, and for each pair FIFA measures the distance between the centre points of the opposing panels.

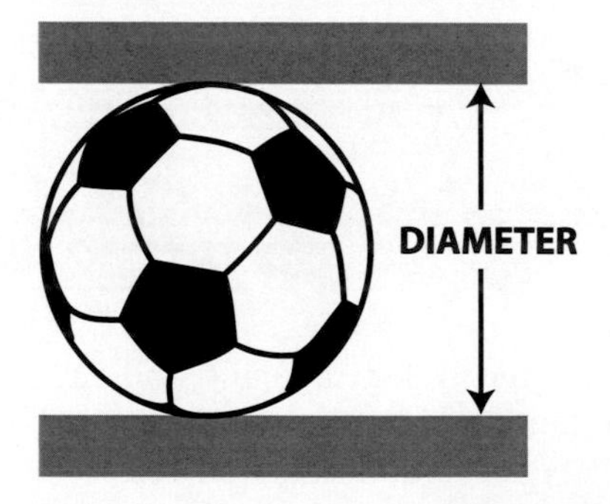

This procedure gives sixteen "diameters" of the soccer ball, and ideally all these diameters would be equal. In practice, FIFA calculates the average of the sixteen diameters, and then declares the ball to be "round" if no individual diameter differs from the average by more than 2%.

The roundness test raises an interesting question: if all sixteen diameters are the same does that guarantee that your "ball" is perfectly round? The answer is clearly no: sixteen equal diameters can determine, at most, the location of thirty-two points around a centre; then we can easily fill in the spaces between those points with flat polygons, and the resulting polyhedron will be perfect according to FIFA's test.

So, what if FIFA decides to be super careful and to test the infinitely many possible diameters, in all directions? Does this finally guarantee that the soccer ball is perfectly round? Again, the answer is no.

We have already written about shapes of constant diameter that are not circles.[2] An example is the rounded blue "triangle" illustrated below. Now, spin this triangle on its bottom point, like a top. The result is a solid shape of constant diameter: it will be judged perfect, even by the super-careful, infinite version of the FIFA test.

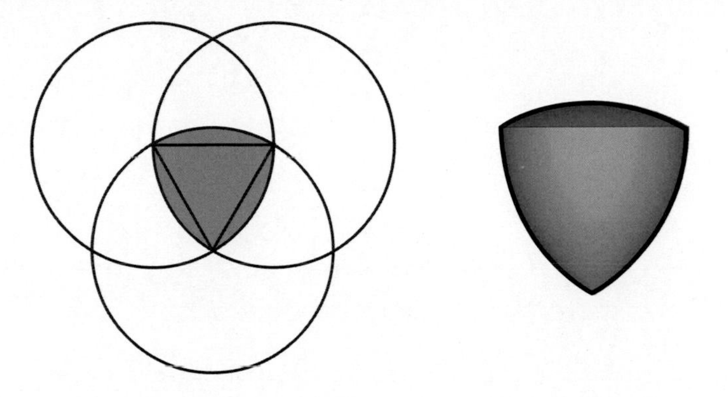

Does this mean that FIFA is using a silly test? No, not really. But there are indeed more reliable methods of testing roundness than simply to measure a couple of diameters. And in situations that are more life-and-death than a soccer game, diameter tests should be avoided.

One such case, where the life-and-death was all too real, was the Columbia space shuttle disaster. The inquiry following the disaster uncovered a number of serious shortcomings in the running of the shuttle program. One shortcoming, although not one responsible for the disaster, was a test for roundness of rocket parts: NASA's test consisted of nothing more than measuring and comparing three diameters.

Puzzle to ponder

Suppose you have a mystery object, and suppose that its shadow (with the Sun directly above) is a perfect circle. You rotate the object and still the shadow is a perfect circle, and you rotate once more with the same result. Do the three circular shadows guarantee that your mystery object is a perfect sphere?

[2]See Chapter 39 of *A Dingo Ate My Math Book.*

CHAPTER 32

And the winner is ...

It's so exciting! The Winter Olympics began last week, and we've been watching intently, working hard to choose a winner. No, not the winner of any of the events. We're looking to award a prize to the Olympic event with the stupidest scoring system.

A few years back this would have been easy, with figure skating winning hands down. You may in fact recall some of the odd happenings at the 2002 Winter Olympics. The most bizarre incident occurred in the women's singles figure skating. In that event the American skater Michelle Kwan was in the gold medal position with only one skater to go. The final skater was the Russian Irina Slutskaya, and she skated extremely well. As a consequence, Michelle dropped from first to ... third.

How could one final skater result in Michelle dropping from gold to bronze? Here's how. Each contestant skated twice, in a short program and a long program. In each program the skaters were scored, and the scores were used to rank the skaters. That's fine. The bizarreness comes with trying to combine the rankings for the two programs.

Common sense suggests that we simply forget the rankings, suitably weight the scores for the short and long programs, and add those weighted scores; then the best total score wins the gold. Instead, the 2002 system involved treating the rankings as numbers, and then those numbers were added, as indicated in the table below. As a consequence, a skater didn't know her ranking number for a program until the last skater had skated.

Lower ranking numbers indicate better performances, so we see Michelle Kwan was in the lead after the short program. The long program is more important, which is reflected in the doubled numbers. In the long program Sarah Hughes skated better than Michelle but it was not enough, so it seemed, to overtake her.

	short program	before Irina		after Irina	
		long program	total	long program	total
Michelle Kwan	0.5	2.0	2.5	3.0	3.5
Sarah Hughes	2.0	1.0	3.0	1.0	3.0
Sasha Cohen	1.5	3.0	4.5	4.0	5.5
Irina Slutskaya	1.0			2.0	3.0

But then Irina skated, finishing second in the long program. As a consequence, Irina lowered Michelle's ranking number, overtook Michelle for silver and, on a tiebreak, propelled Sarah Hughes into the gold medal spot. What a performance.

Alas for absurdity fans, the International Skating Union modified their rules in 2004, and rankings are no longer added together. So, figure skating is out of the running in this year's Stupid Scoring competition. But there are still some worthy entrants.

We might consider sports such as the bobsled and the luge, but they clearly have no chance: these events amount to turning yourself into a missile, with the missile taking the least time to get to the bottom of the hill winning the gold. Many sporting events are of this nature, with some objective physical quantity determining the winner.

There are also many sports with a subjective element, such as mogul skiing and, of course, figure skating. More intriguing are sports such as the biathlon, consisting of skiing and rifle shooting; performance in each sport can be precisely evaluated, but there is unavoidable subjectivity in weighting the two performances. We can't imagine how the authorities determine the weighting.

However, it is a trifle unfair to make fun of sports with a natural subjective component. We would much rather make fun of a sport that is judged subjectively, even though an objective measure is right at hand. But who would possibly do such a thing?

Welcome to ski jumping. This magnificent event is the highlight of the Winter Olympics, where skiers shoot off the ramp at close to 100 km/h and fly more than 100 meters through the air. And of course, it's easy to determine the winner: whoever jumps the furthest gets the gold, right? Wrong.

In its wisdom, the International Ski Federation has decided that ski jumpers should be judged not only on distance, but also on "style". But it is simply beyond us how anyone can regard the "best style" as anything other than the one which takes you the furthest. And, we wonder how the Federation explained it to the Olympic silver medalists in 2006: both Matti Hautamaeki and Andreas Kofler would have won gold based upon distance scores alone.

With their stunning entry, the International Ski Federation has blitzed the Stupid Scoring competition. Ladies and Gentlemen, we have a winner.

Puzzle to ponder

With the 2002 scoring system, Michelle was sitting first with one skater to go, and fell to third, but at least she still went home with a medal. Is it possible that a skater could have been coming first with one skater to go, and then fallen to fourth?

Part 4

The House that Math Built

CHAPTER 33

Visionary Voronoi

It's happened again. After six years of cataloguing Victoria's mathematically inspired architecture, your Maths Masters were sure they'd seen it all.[1] But no: in the heart of Melbourne we've stumbled across yet another beautifully mathematical building.

Earlier this year we took an excursion to Melbourne's outskirts, to visit the Klein bottle house on the Mornington Peninsula.[2] Now we're back in the city, checking out the Victorian College of the Arts' Centre for Ideas. Located in Southbank, the award-winning CFI was completed in 2001. It was designed by Melbourne-based Minifie van Schaik Architects, who were also responsible for Healesville Sanctuary's Wildlife Health Centre, the inspiration for another of our mathematical expeditions.[3]

CFI's striking façade is inspired by what is known as a *Voronoi diagram.* We pondered such diagrams a few years back, when discussing the zoning of Melbourne's public schools.[4] The following diagram shows the zoning of secondary schools in the neighborhood of Glen Waverley Secondary College.

---

[1]See Part 4 of *A Dingo Ate My Math Book.*

[2]See Chapter 22 of *Dingo.*

[3]See Chapter 21 of *Dingo.* Alas, the mathematical "minimal surface" roof of the Wildlife Centre has since been replaced by a an ordinary flat roof. It turns out that the "optimal" roof made things hot as Hell; so, not such an optimal roof after all.

[4]See Chapter 42 of *Dingo.*

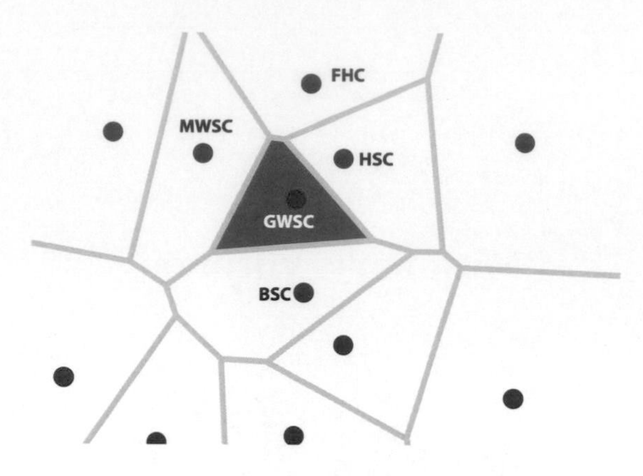

To construct the diagram we begin by plotting the locations of all the schools. Then each school's zone, called a *Voronoi cell*, comprises all those points that are closer to that school than to any other.

In our previous column we noted that all the edges of a Voronoi cell must be straight lines. Moreover, the common edge of two adjacent zones is easily located: draw the straight line connecting the two schools, and then the shared boundary is the perpendicular bisector of the connecting line.

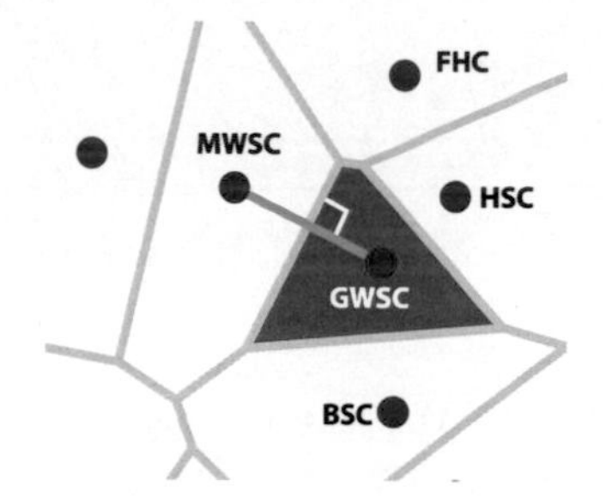

This provides a natural, if somewhat fiddly, method of constructing a Voronoi diagram. There turns out to be a very clever and much prettier method; it is this method that inspired the CFI façade.

Start with the same points/schools as before, as in the left diagram below. Now, imagine each school hovering the same height above the ground, as pictured in perspective on the right.

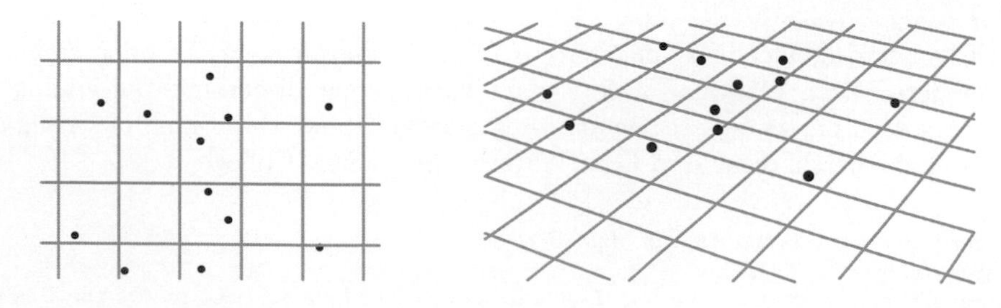

Next, for each school draw a cone with apex at the school and descending down to the ground.

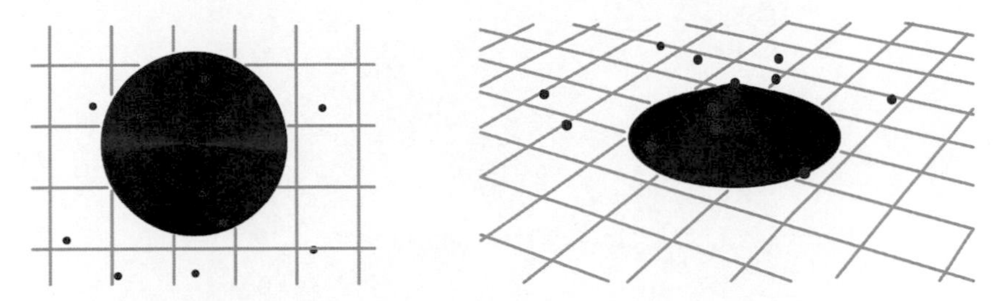

The cones will tend to run into each other, intersecting along (hyperbolic) curves. Any such curve of intersection will be equidistant from the apices of two cones. That means the top view of these curves exhibits the perpendicular bisectors that we used above to construct the Voronoi diagram.

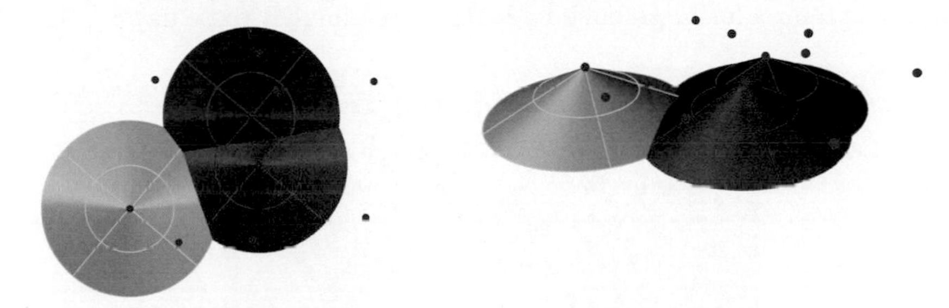

Taking all the cones together, the top view of the resulting landscape exactly gives us our school zoning map. Very cool.

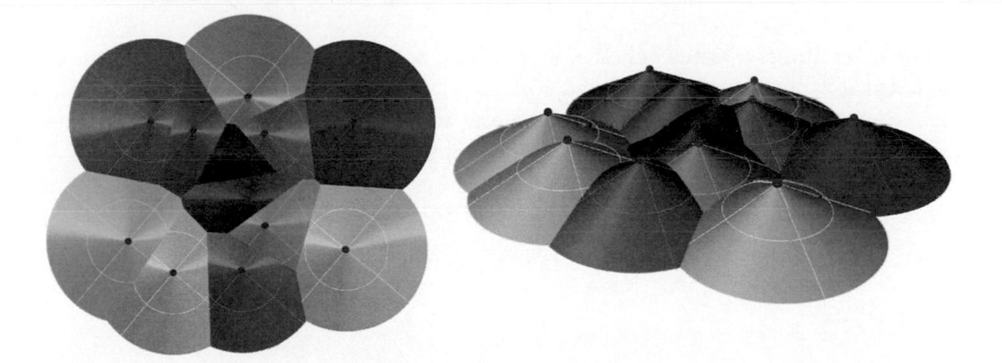

With a little highlighting, it's easy to see the Voronoi diagram in the CFI façade, together with the associated cones, pointing inwards. It is a striking effect, and there is another hidden piece of engineering cunning. Since any part of a cone can be cut from a flat piece of metal the façade would have been relatively easy and cheap to construct.

So, not only are Voronoi diagrams extraordinarily useful, with all manner of important applications,[5] they can also be extraordinarily beautiful. Another happy end to another mathematical architectural tale.

Almost.

Alas, the Centre for Ideas building looks quite different these days:

Little seedlings planted when the building was opened are now all grown up, almost totally obscuring CFI's beautiful façade. We are sure that the Centre for Ideas is a fine institution, the home of many worthy ideas. The planting of those large trees, however, is one idea that they could well have done without.

Puzzle to ponder

Describe the regular pattern underlying the Voronoi tiling of the façade.

[5]Most famously, in 1855 the physician John Snow used a Voronoi diagram to identify the source of a cholera outbreak in Soho, London.

CHAPTER 34

Melbourne's catenary chaos

You just moved into a new house on a busy road, and it seems that you'll never get used to the noise of the thundering trucks. Except that you do.

Melburnians are expert at shutting out the ugliness of city living. (Melbourne's great food and coffee helps.) And there is one form of visual pollution that we all work very hard to ignore: the ludicrous tangle of cables and power lines suspended from the ossified jungle of dead trees.

The fine product of third world planning, Melbourne's wiry mess is truly grotesque. So, it is with sincere apologies that we now ask you to focus upon this ugliness.

We want to ponder the shape of the curves made by all those cables. Here is a very slack example, anchored at two level points.

It may look familiar: is Melbourne suffering from parabolic pollution, and is this finally justification for inflicting all that quadratic torture on school students?[1] Nope.

For comparison, over the blue cable we have superimposed a red parabola, which also passes through the two anchor points as well as the bottom point of the cable. The curves pictured look similar but they are not identical.

So, if hanging cables are not parabolas then what are they? Actually, they're not a "they", they're an"it".

Though hanging cables appear to come in various shapes, there is a sense in which they are all just portions of one master curve, the so-called *catenary.* ("Catena" is the Latin word for chain). The blue cable above is the bottom part of the catenary.

How can we identify all hanging cables? First hammer two nails into your classroom wall. (Don't bother asking for permission from your teacher; we're sure she won't mind.) Then hang a thin cable or chain between the nails, forming the blue curve. Next, picture all of the scaled up versions of the blue curve and shift them to pass through the nails, as shown.

What we are claiming is that this picture represents all the shorter ways of hanging a cable between the two nails. (To get the longer, droopier cables, we have to imagine beginning with a longer blue cable.)

[1]See Chapter 5.

How can we see this? To begin, notice that the shape of a hanging cable does not depend upon its weight. For example, having a cable of twice the weight would clearly take the same shape as two of the original cables hanging side by side.

It's a little tricky, but this independence from weight shows that a scaled image of a hanging cable again represents a hanging cable. Instead of a uniform cable, imagine a string of pearls; then, simply by looking at the forces acting on each pearl, it is clear that scaling up the string of pearls will hang in exactly the same manner as the original string. The same must then be true for uniform chains.

It is also the case that any part of a hanging cable is still a hanging cable. That is, if you imagine pinching your fingers around a cable at two points, and let the pieces of chain above your fingers go loose, the cable below your fingers won't change shape.

Together, these observations prove our claim, that any hanging cable is just a scaled portion of one master curve, our catenary. But exactly what shape is the catenary? To determine the precise shape takes work, and either a little calculus or a lot of cleverness. One approach is to grab another pearl necklace, with many small pearls, connected by short and essentially weightless pieces of string. Then, one can carefully analyze the forces at work.

It turns out that the catenary can be written easily in terms of the exponential function, involving the fundamental mathematical constant e. (The number e is often erroneously referred to as "Euler's number", but that's another story.[2]) University students know of the catenary as the *hyperbolic cosine function* (see the puzzle), although they are seldom told why this just-another-function is at all interesting.

As well as describing hanging cables, the catenary is famous for another striking appearance in nature – the soap film that forms between two parallel circular wire loops. This soap film, which takes the shape of a rotated catenary, is known as a catenoid.

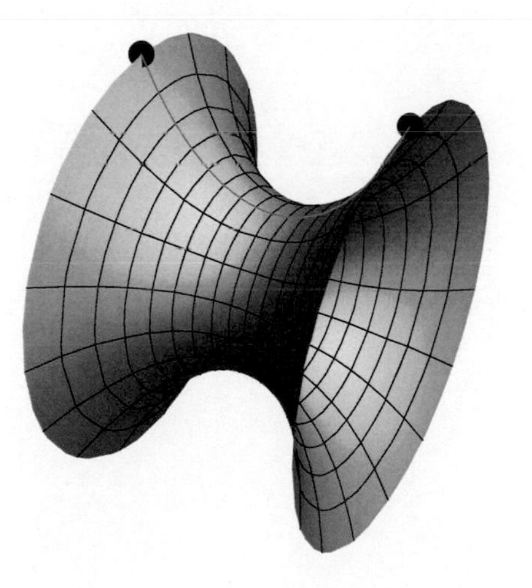

[2]See Chapter 7.

And here is a final, eccentric application of the catenary. Imagine constructing a catenary-shaped floor as in the picture below. Then, a square wheel of just the right size can roll smoothly over the floor. And a smooth ride on a square wheeled bicycle is not just a theoretical possibility: it can be an amusing reality.[3]

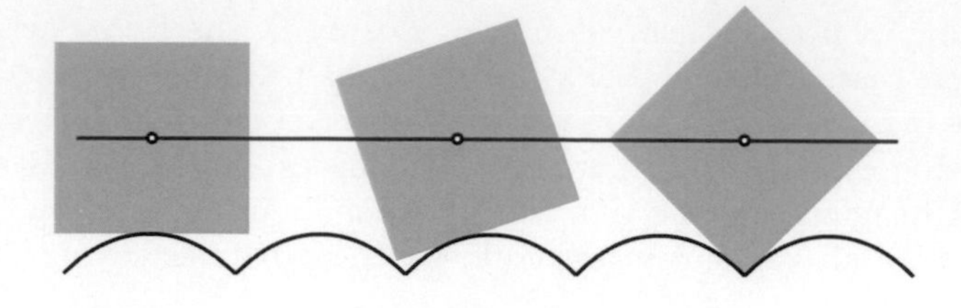

Puzzle to ponder

The catenary has equation $y = \frac{e^x + e^{-x}}{2}$. Using this, and assuming that the rolling square has sidelength 1, what is the length of the waves in the wavy floor?

[3]Videos of such bikes are easily found online.

CHAPTER 35

Eureka!

Most Melburnians would be aware that our skyscraping Eureka Tower is named after the Eureka Stockade, the iconic battle between miners and the military from Ballarat's gold rush days. But why "Eureka"?

The Greek word Eureka means "I have found (it)", a very handy expression for a time and place where people were stumbling upon nuggets of gold. But why would an Australian miner exclaim his delight in Greek? It all stems from the granddaddy of Eureka moments, which also involved a quantity of gold.

Archimedes, the mathematical superstar from ancient Greece, was presented the task of determining whether a certain crown was solid gold. To do this, Archimedes only needed to calculate the volume of the crown, preferably without destroying it. While taking a bath, he suddenly realized that the crown's volume could be determined by placing it in water and measuring the volume of water displaced. Thrilled with his discovery, Archimedes jumped up and ran naked through the street, shouting his famous cry.

It's a great story, and possibly even true. However, we want to discuss a different Eureka moment, definitely true, and grand enough to have been commemorated on Archimedes' tomb.

Elsewhere, we wrote about the mathematics of ancient Egypt,[1] and we raised the question of whether the ancient Egyptians knew of the formula for the surface area of a sphere. It is unlikely, but in any case Archimedes *proved* the formula, something the Egyptians most definitely did not do. A reader subsequently wrote to us, asking how Archimedes accomplished this.

We endeavor to keep our readers happy, but this story will take some time to tell. To determine the area of a sphere, Archimedes first had to know the formula for the *volume* of a sphere. So, today we'll discuss Archimedes' ingenious study of volumes. Surface areas will be the Maths Masters' homework for next week.[2]

We learn in school that a sphere of radius R has a volume of $\frac{4}{3}\pi R^3$, but seldom does anyone ever hint at why this formula works. How could Archimedes prove such a formula?

To begin, we have to discuss yet another instance of Archimedean brilliance: *the law of the lever.* This law states that two objects on a scale will balance if the weight multiplied by the distance to the fulcrum is the same for the two objects. For example, on the scale pictured below, the 100 kg globe will exactly balance the 60 kg Archimedes, because $100 \times 3 = 60 \times 5$.

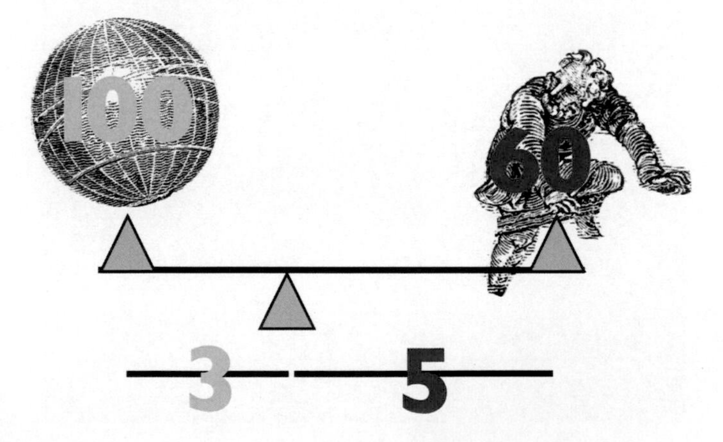

Now, what to do with this law? As well as the incredibly important practical applications, Archimedes had a new way of playing with geometric shapes. So, he went to his toy box and grabbed a sphere of radius R; it was the volume of this sphere that Archimedes wanted to find. Archimedes also fetched a cylinder and a cone, both with height $2R$ and circular base of radius $2R$.

[1] A column we wrote, mostly on what are known as Egyptian fractions.

[2] See the next Chapter.

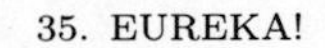

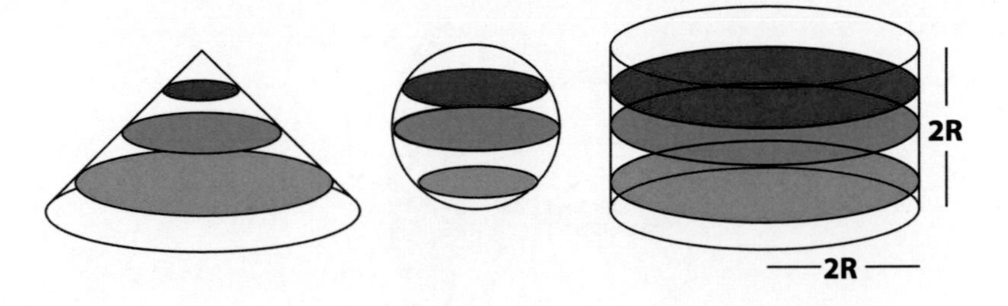

It is easy to calculate the volume of the cylinder ("base times height"), and Archimedes also knew that the volume of the cone was one third that of the cylinder: the mathematician Eudoxus had worked that out about a hundred years earlier. Archimedes' brilliance was to recognize that his three solids could be balanced on the scale, as pictured:

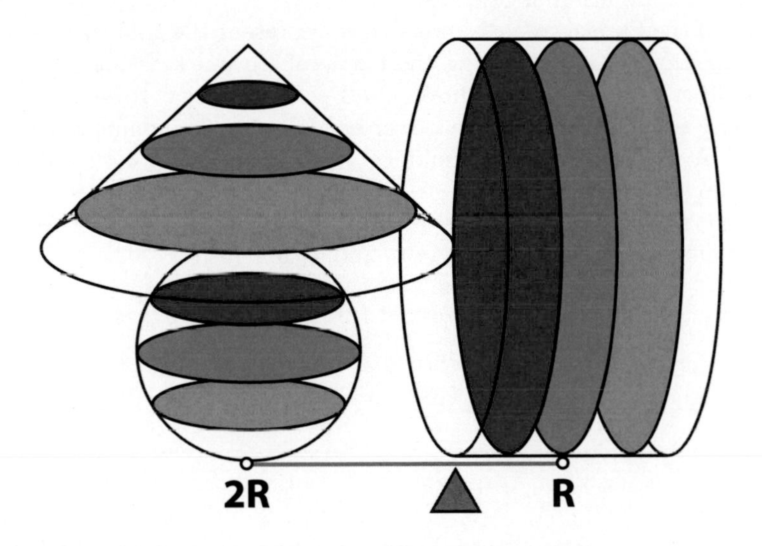

From the law of the lever, Archimedes then knew that

$$2R \times (\text{volume of cone} + \text{volume of sphere}) = R \times (\text{volume of cylinder}).$$

So, knowing the volumes of the cone and the cylinder, some simple arithmetic gave Archimedes his desired result, the (now) well-known formula,

$$\text{volume of sphere} = \frac{4}{3}\pi R^3 .$$

But does balancing solids on a scale amount to a proof? The truly ingenious part of Archimedes' balancing act is indicated by the pictured slices of the solids. The three solids have the same height, and the slices at corresponding levels are indicated by the colors.

Archimedes noted that any two slices of the same color on the left are balanced by the slice of the same color on the right. And, it is straight-forward to *prove* that this is the case, because we are simply comparing the areas of three circles. Archimedes then argued that because the slices of the figures exactly balance, the whole solids must balance as well.

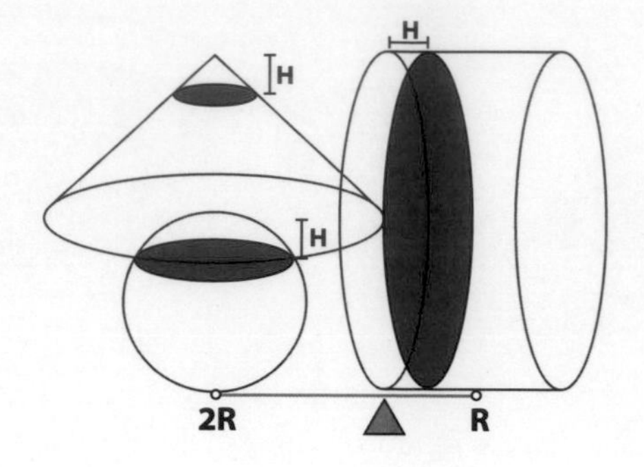

Archimedes' genius here is reminiscent of the fundamental idea behind calculus, which was only properly developed 1900 years later. Indeed, many have argued that Archimedes (and, for similar reasons, Eudoxus) should be given at least some of the credit for the invention of calculus.

The top of the Eureka tower is intended to represent the gold rush, and the red line the blood that was shed during the battle of Eureka stockade. But, without Archimedes there would be no Eureka, and so no Eureka Tower. Your Maths Masters prefer to regard the Eureka Tower as Melbourne's monument to the great Archimedes. And the red line? It could signify Archimedes' pointless death at the sword of a thuggish Roman soldier.

And, what about the surface area of a sphere? That is yet another Archimedean Eureka moment. As promised, we'll write about that next.

Puzzle to ponder

Can you provide the details for the last diagram, to show that the two left circles balance the right circle? (Another Greek mathematical hero, Pythagoras, may be of assistance.)

CHAPTER 36

Archimedes' crocodile

Our previous column stopped halfway through a great mathematical story.[1] We wrote of Archimedes' Eureka moments and his beautiful derivation of the volume of a sphere. We concluded with a promise to follow up with his equally beautiful derivation of the surface area of a sphere.

We'll do that, but we first want to introduce you to Archimedes' crocodile. Actually, we don't know who found him originally, but the crocodile did eventually become Archimedes' pet.

We've previously discussed a version of the crocodile.[2] Take a pizza and chop it into an even number of equal slices.

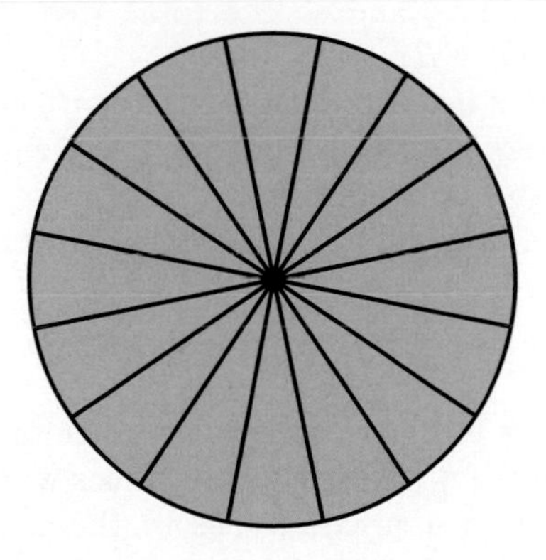

[1]See the previous Chapter.

[2]See Chapter 48 of *A Dingo Ate My Math Book* , where we used a cooking scenario to present some mathematical gems.

Then, rearrange the slices into the big toothy smile of a crocodile.

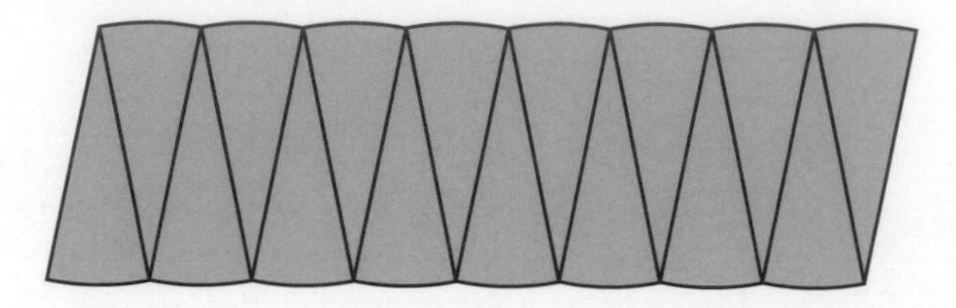

How big is this smile? If the pizza has been sliced into a large number of thin pieces then the smile is very close to being a rectangle. The height of the "rectangle" will be close to the radius R of the pizza, and the base will be half its circumference.

We can then see at a glance that

$$\text{Area of the pizza} = \text{Area of the smile} = 1/2 \times (\text{Circumference of the circle}) \times R\,.$$

A slightly different method of obtaining this equation is to recognize that the area of each (roughly) triangular tooth equals $1/2 \times \text{base} \times R$. Adding the bases, they sum to the circumference, and so the total area of all the teeth is $1/2 \times$ (Circumference of the circle) $\times R$.

As a consequence of this beautiful equation, if you know the area of a circle of radius R then you can determine its circumference, and vice versa. At which point, the familiar π enters the story.

What exactly is π? It's actually a little tricky, but the easiest approach is to think of it as the precise number that makes true the following famous equation:

$$\text{Area of a circle} = \pi R^2\,.$$

The hidden trickiness is in knowing that the exact same number will work for every circle. But to continue, by approximating circles with polygons, and by other methods, we can then obtain approximations to the size of the number π: the familiar 22/7 and 3.14, and so on.

Now, Archimedes' crocodile has a message for us: since we have an exact formula for the area of a circle, we also have an exact formula for the circumference of a circle: rearranging the above smile equation, we find that

$$\text{Circumference of a circle} = 2\pi R\,.$$

Waving goodbye to Archimedes' crocodile, we now return to pondering spheres. As we noted, Archimedes knew that the volume of a sphere of radius R is $\frac{4}{3}\pi R^3$.

But then Archimedes realized that the exact same pizza-crocodile trick works to determine the surface area of the sphere.

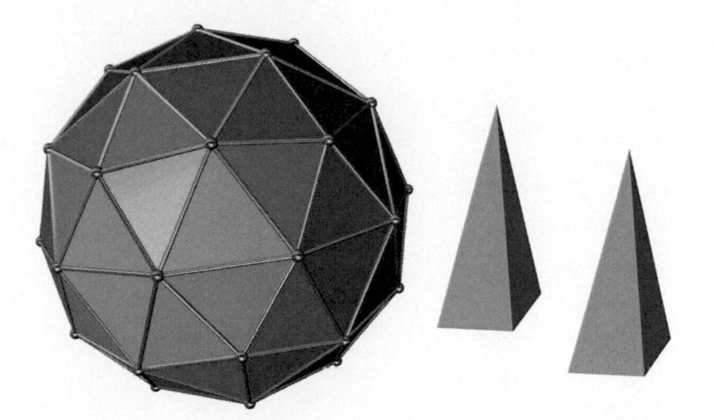

We simply have to think of the sphere as consisting of a very large number of thin pyramidal teeth. As we mentioned last week, each tooth will have a volume of $1/3 \times$ (area of the base) $\times R$. And, summing the bases of all the teeth gives the surface area of the sphere. So, summing the volume of the teeth,

$$\text{Volume of sphere} = \frac{1}{3} \times (\text{sum of the bases}) \times R = \frac{1}{3} \times (\text{Surface area of sphere}) \times R$$

Rearranging, we find, as Archimedes did,

$$\text{Surface area of a sphere} = 4\pi R^2 .$$

Our final, crocodile-inspired, Eureka moment.

Puzzle to ponder

Can you indicate why π is the same for every circle? To make it clearer, let's define "π" to be the area of a circle of radius 1. How, then, do you know that a circle of some other radius, say R, has area πR^2?

CHAPTER 37

Melbourne Grammar mystery map

On a recent visit to Melbourne's Shrine of Remembrance we stumbled across another beautiful shrine: for mathematics. This is the intriguing sculpture at the entrance to Melbourne Grammar School. It is constructed from equilateral triangles and, somehow, is a map of the Earth – Australia and New Zealand are clearly visible in the top right corner. We want to explain what this strange map is all about.

People have been making maps of the Earth for thousands of years and it has always been frustrating, at least since people realized the Earth is Earth-shaped. One would hope that by now we would have devised the perfect map. On such a perfect map countries of equal area would also have equal area on the map, the shortest path between two points would be represented by a straight line, right angles would appear as right angles, and so on.

Alas, there is no perfect map of the Earth, for two quite distinct reasons. First of all there is the *topology* of the Earth: the spherical form of the Earth, with no cliffs at an "edge" to fall over, ensures that certain paths, such as the equator, inevitably loop back on themselves. We cannot faithfully represent all such loops on a map: the only alternative is to rip the Earth apart and to map the tattered remains.

But what if we just wanted to map a small piece of the Earth? Here, the second problem, *geometry* – literally, the measurement of the Earth – rears its head. The great 19th century mathematician Karl Friedrich Gauss showed that such a perfect map cannot exist, even for a tiny piece of the Earth.

To see Gauss's argument in action, imagine we want a map of the Arctic. Starting at the North Pole, we could go south a given distance in all directions, marking off a circular region. On a perfect map this region would also appear as a circle, and smaller circles of radius R within the map would have an area of πR^2. However, this simply *must* be a distortion of the actual Arctic, since the tried and true formula AREA $= \pi R^2$ is *not* true for circles around the North Pole.[1] So, if the diameters of the circles are correctly scaled then the areas must be wrong, and vice versa. It is in such measures that the geometry of the Earth is fundamentally different from that of flat maps.

So, maps of even small portions of the Earth are doomed to distortion. As a necessary compromise, mapmakers focus on creating specialized maps, serving designated functions extremely well. Examples are aeronautical maps, which preserve important concepts of direction, and the maps commonly found in classrooms, which are generally good compromises and not too distorted in any given way.

Now, back to the topology problem. We still have to decide how to rip the Earth apart, and our mystery sculpture encapsulates one such approach. The idea is to first approximate the spherical Earth by a shape with flat sides. We then cut the shape open along some of its edges and flatten it out, resulting in our map. The shapes that come to mind are the Famous Five: the regular solids pictured below.

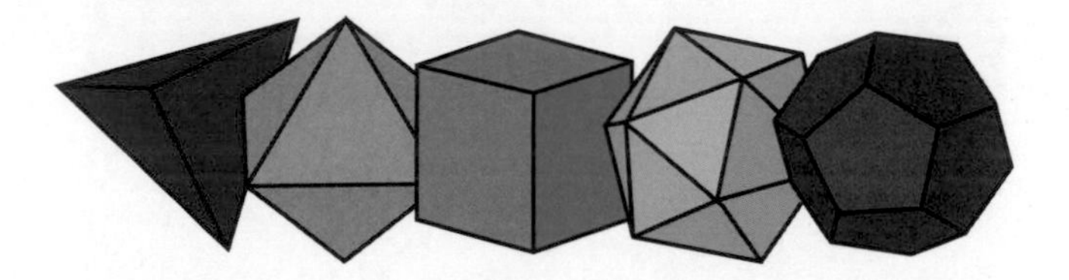

Suppose we choose the cube. Imagining the surface of the Earth just fitting in the cube, a light at the centre of the Earth would then project an image onto the cube. Now cut the cube open and flatten, giving you a map of the Earth. The resulting map is surprisingly similar to more familiar maps.

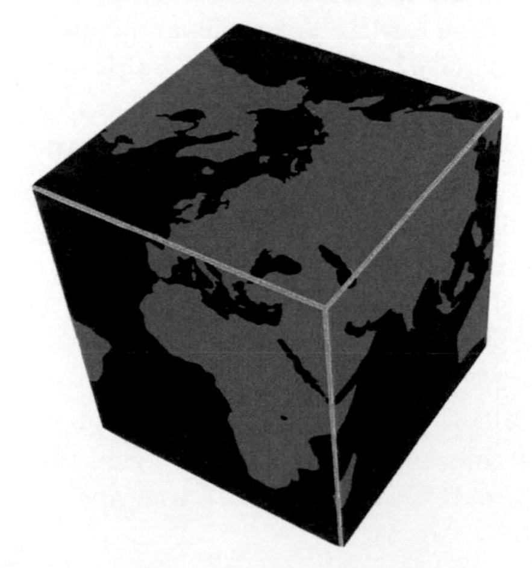

[1]With a little calculus, it is not difficult to show that the area of such a spherical circle of radius R is $2\pi E^2\left(1-\cos\frac{R}{E}\right)$, where E is the Earth's radius. If R is not too large then $\cos\frac{R}{E}\approx 1-\frac{R^2}{2E^2}+\frac{R^4}{24E^4}$. That gives the area of the circle as approximately $\pi R^2-\pi\frac{R^4}{12E^2}$, a little less than an everyday flat circle of the same radius R.

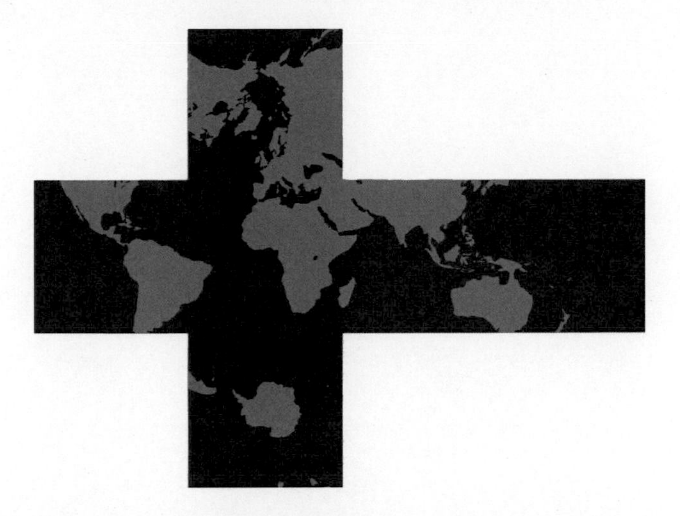

One can do better, by starting with the icosahedron, the regular blue solid pictured above. With twenty small triangles, the icosahedron hugs a sphere much closer than a cube. Choosing a certain projection, and making the right cuts, you can unfold to give our beautiful sculpture.[2]

The idea behind such flat projections dates to the 15th century painter Albrecht Dürer. The icosahedron map itself was inspired by Buckminster Fuller, as one of his Dymaxion maps, created in 1954. And, complete disclosure, the Melbourne Grammar sculpture is missing three triangles from the icosahedron map. We suspect that some less important continent may have been sacrificed for the sake of structural soundness.

Puzzle to ponder

A bear leaves his home. He walks 20 miles South, then 20 miles East, then 20 miles North. He discovers that he is home again. What color is the bear? What if the bear were a bird? What color would he be then?

[2]In the original article we used some wonderful maps created by Carlos A. Futuri, but unfortunately Futuri's maps appear to be no longer available.

CHAPTER 38

A rectangle, some spheres and lots of triangles

In another column, we left you in suspense with an unanswered question: how do we make a mathematically perfect rubber band ball? We answer that below, but we'll first digress and take a close look at the bubble dome, Melbourne's striking new sports stadium.

Known officially as the Rectangular Stadium or AAMI Park, the bubble dome consists of pieces of what are approximately *geodesic spheres*. In turn, a geodesic sphere is a network of small triangles. (Well, in the case of the bubble dome, "small" means having sides about three meters in length). Geodesic spheres became very popular in the 1960s, as an alternative model to conventional rectangular architecture.

Why would anybody want a spherical house? Well, among all shapes enclosing a given volume, the sphere has the least surface area. So, a spherical house requires less building material and the smaller wall area would potentially result in less heat loss. Geodesic spheres are also light and very stable. True, there are also inefficiencies. It's not easy to make good use of spherical walls, those edges of the many triangles have to be sealed, and so on. So, though many geodesic homes have been built, there are definite shortcomings. However, it is very natural to employ a geodesic roof to shelter the rising tiers of a sports stadium. It also looks really cool.

Some impressive examples of geodesic constructions are the Montreal Biosphere and the Nagoya Dome.

A traditional starting point for a geodesic sphere is the *icosahedron*, one of the famous five Platonic solids. We've had occasion to mention these before, when designing silly soccer balls.[1] An icosahedron has 12 corners, 30 edges and 20 equilateral triangular sides. The corners all lie on a sphere, which the icosahedron roughly approximates. So, the icosahedron can be regarded as a primitive geodesic sphere. To obtain something more refined, we can subdivide each triangle of the icosahedron into smaller equilateral triangles, and then push the new corners outwards until they also lie on the sphere. The more we subdivide, the better the approximation to the true sphere, as illustrated by the two geodesic spheres below.

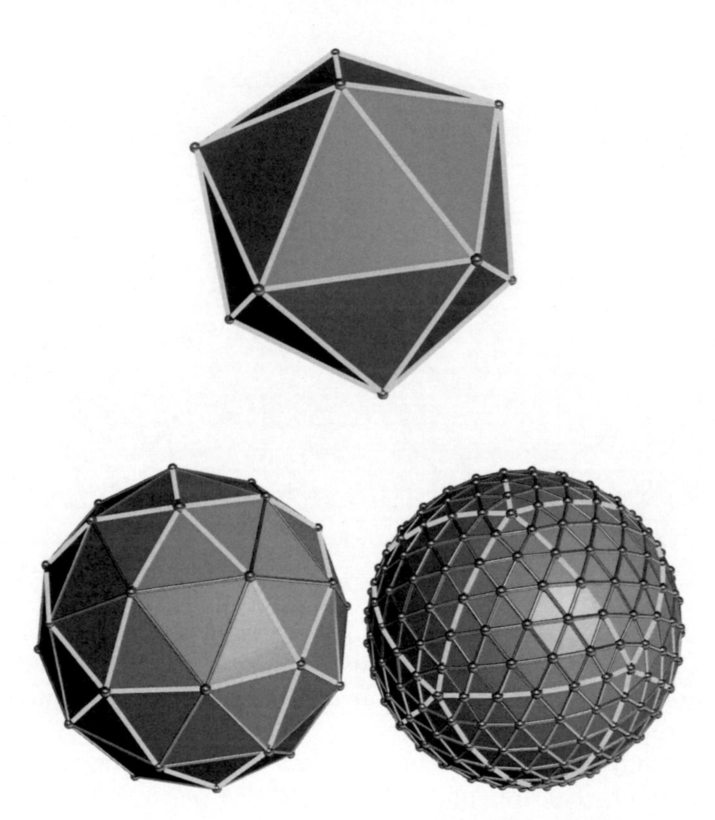

[1]See Chapter 31.

Architectural wizardry aside, your Maths Masters have recently discovered a beautiful new use for geodesic spheres. They can be used to create ideal, mathematically perfect cousins of messy rubber band balls.

In a standard rubber-band ball, the bands are included as great circles placed any which way. To make the construction less arbitrary, we want a collection of circles that is evenly and symmetrically distributed, and that's where the geodesic sphere comes in.

The corners of a geodesic sphere come in diametrically opposite *antipodal pairs*, just like the North and South Poles on the Earth. Moreover, just as there is the Equator equidistant from the poles, each antipodal pair of corners has an associated *great circle*. The corners are also very evenly distributed, so if we draw a great circle for each pair of antipodal corners, we should obtain an impressively symmetric model for a rubber band ball. And we do! (Well, there is still the issue how the rubber bands would overlap each other, but we'll leave those technicalities for the Construction Department.)

Consider the two geodesic spheres we constructed earlier. In what may amount to a world premiere, below are the two corresponding rubber band balls. Enjoy!

Puzzle to ponder

How many rubber bands (equators) are on each of the two spheres above?

Part 5

A Chance Encounter

CHAPTER 39

So you think you can beat the bookies?

In this era of economic (pseudo) rationalism, you've got to admire a city that takes a day off for a horse race.

We love the Melbourne Cup.[1] We don't love horseracing, but the general frivolity warms our hearts. And, we're intrigued by its much less frivolous heart: the mathematics of gambling.

Is it possible to make money, *reliably*, gambling on horse races? Yes, of course it is, and it requires very little mathematics: you simply have to be a very good judge of horseflesh. There are gamblers who succeed at this, but it is hard work. You need to know the horses *very* well, well enough to overcome the fact that the payouts are rigged against you.

Essentially all gambling games are rigged, so that the company offering the game is guaranteed a profit. That's not a surprise, and it's hardly unreasonable. What is much less reasonable is that the extent of the rigging is rarely made explicit. We suspect that few punters have any real sense of the rigging.

---

[1] Yep, don't ask us why, but the first Tuesday of November, the date of the Melbourne Cup, is a public holiday in the state of Victoria.

Luckily, there's a simple method to work this out. If you bet just the right amounts on each possible outcome then, no matter who wins, you'll receive the same sum back. Then, the rigging is indicated by comparing the total amount of money you've wagered to the sum returned.

To see the method in practice, suppose you want to bet on the upcoming Ashes cricket series.[2] At the time of writing, one betting company is offering the following prices:

Australia	**\$1.65**
England	**\$3.30**
Draw	**\$5.50**

The table indicates, for example, that a winning \$100 bet on England will return \$330, for a profit of \$230. Now, you may regard that \$3.30 as good odds against a less than impressive Australia: maybe the betting company has it wrong. But, independent of judging cricketflesh, you can measure the extent to which the odds are intrinsically rigged. To do this, you simply *take the reciprocals of all the payouts, and add.* For our Ashes example, the calculation is

$$\frac{1}{1.65} + \frac{1}{3.30} + \frac{1}{5.50} = 1.09\,.$$

The final 1.09 indicates that betting a total of \$109 on the three possible outcomes – in just the right proportions – will guarantee you a return of exactly \$100, no matter how the series unfolds. Then, the extra \$9 you are required to wager indicates the extent of the rigging.

We can now apply exactly the same method to consider potential bets on the Melbourne Cup. On the next page is the table of payouts offered (at the time of writing) by a number of online betting sites.

The last row of the table indicates, for each of the betting companies, the amount to wager to guarantee a return of \$100. Not surprisingly, this amount is always greater than \$100, and none of the sites is offering enticing odds. Company D is the best of an unattractive bunch, requiring \$115 to be bet for that return of \$100.

So, what should you do? If you have simply decided on the horse you fancy, then you just choose the best site for that horse: Company D for *Maluckyday*, Company C for *Americain*, and so on.

On the other hand, if you want to reduce the overall rigging, you can do so by choosing the best site for each horse. Indeed, if you're lucky, the rigging may disappear entirely, guaranteeing a \$100 return on *less* than \$100 wagered. This is a famous approach to gambling, known as *arbitrage betting.*

In the last column of our table, we've identified the best payout for each horse, and the final entry in the column indicates the overall rigging. As is indicated, you still need to lay out \$110 to guarantee a \$100 return: still not at all enticing.

[2]Dating from 1882, the Ashes is the biennial cricket contest between England and Australia. The contest typically consists of five "tests" (matches), each test lasting five days, a test often enough ending in a draw. It's very weird, and a big deal.

	A	B	C	D	E	F	BEST ODDS
SO YOU THINK	3.4	3.4	3.25	3.4	3.4	3.4	3.4
SHOCKING	9	10	9	10	9	10	10
MALUCKYDAY	9.5	9	9.5	10	9	9	10
AMERICAIN	11	12	14	13	11	12	14
DESCARADO	12	12	14	13	13	12	14
MANIGHAR	16	19	19	19	17	21	21
HOLBERG	18	17	21	23	17	19	23
PRECEDENCE	19	17	19	18	21	19	21
LINTON	21	17	18	26	21	17	26
HARRIS TWEED	26	26	21	23	23	26	26
MONACO CONSUL	21	31	26	23	23	31	31
PROFOUND BEAUTY	26	31	21	26	26	31	31
ZIPPING	31	31	26	31	31	31	31
SHOOT OUT	31	31	26	31	31	31	31
ONCE WERE WILD	34	26	31	35	23	26	35
MR MEDICI	51	51	51	51	51	51	51
CAMPANOLOGIST	61	41	67	61	61	41	67
ILLUSTRIOUS BLUE	61	61	51	81	61	81	81
BAUER	61	81	61	61	71	81	81
TOKAI TRICK	71	81	81	81	81	81	81
ZAVITE	126	151	126	126	126	151	151
MASTER OREILLY	201	301	151	201	201	301	301
BUCCELLATI	301	201	251	251	251	251	301
RED RULER	301	301	251	201	301	301	301
BET AMOUNT	123	121	122	115	123	118	110

And, after all this, how will your carefully calculating Maths Masters be betting on the Melbourne Cup? Well, we plan to slap a few dollars down on *Profound Beauty*. It's a great name: a guarantee, of course, that the horse will do well.[3]

Puzzles to ponder

Prove our reciprocal formula above. So, suppose the payouts on a three-horse race are $\$A$, $\$B$ and $\$C$. Show that if you bet a total on the horses of $\$(1/A+1/B+1/C)$, in just the right manner, you will be guaranteed a return of $1.

Suppose now a fourth horse, *Son of Seabiscuit*, enters the race, and the odds given on this champion horse are $100. Assuming the odds on the other horses remain the same, has the rigging on the race gone up or down?

[3]It finished 16th.

CHAPTER 40

A Penney for your thoughts

During the school holidays one of your Maths Masters decided to take his junior maths mistress to the zoo. Alas, it was raining cats and dogs. So, we changed course and headed for Scienceworks, Melbourne's hands-on science museum. It was a zoo.

It turns out than on rainy holidays every one of Melbourne's ten billion children ends up at Scienceworks. Which means that despite the best efforts of the beleaguered staff it takes a drizzly hour just to get inside. Then, once inside, it's chaos, a density of ten million children per exhibit.

Of course even if the parents were frazzled the children were all as happy as could be. They rushed excitedly from exhibit to exhibit, seemingly learning at a hundred kilometers per hour, and at a hundred decibels. Moreover, as a surprise maths masterish treat, it turned out that there was a dedicated mathematics section up and running.

Mathamazing is a travelling exhibition put together by Questacon, Canberra's very impressive science museum. It consists of twenty-two hands-on exhibits, capturing some surprising and beautiful aspects of mathematics. Each exhibit is accompanied by a detailed description, indicating what is being demonstrated.

So, a great thing. Kids having lots of fun and learning some math in the process. Except, your Maths Masters are not convinced that more than a handful of children were learning much of anything.

Sure, there were lots of buttons being pushed and plenty of things being banged. But what did the children learn? What did they understand of what they were pushing and banging, or of the accompanying explanations?

Consider, for example, Penney's Game, the subject of one Mathamazing exhibit. The creation of mathematician Walter Penney, the game consists of two players choosing a sequence of three coin tosses, such as TTT or HTT. A coin is then tossed repeatedly and the player whose sequence appears first is the winner. For example, if the sequence of tosses was H H H T H H T H T T then HTT would have won against TTT.

On the face of it Penney's game appears to be pretty boring. If a coin is tossed three times then there are eight possible sequences, each with a 1/8 chance of occurring. So, winning Penney's game would appear to just come down to luck. Surprisingly, that is not the case. For example, it turns out that HTT will beat TTT 7/8 of the time.

In order to understand this, let's determine how TTT might win against HTT. Obviously if the tossing sequence begins T T T, which has a 1/8 chance of occurring, then TTT wins immediately. That turns out be the *only* way, however, that TTT can beat HTT: after a Heads has appeared, the sequence HTT *must* occur before TTT.

So, it turns out that in Penney's game some sequences have a better chance against others. Which means it can sometimes be an advantage for a player to choose second, choosing their sequence in response to the first player's choice. What is amazing is that it is *always* an advantage to go second: no matter the sequence the first player chooses, the second player can always choose a sequence with a better chance of winning.

That seems plain wrong. Surely the first player could simply choose the best sequence and the second player would be stuck with second best. But, it turns out that in Penney's game there is no best choice.

The diagram below indicates for each sequence a second sequence that will beat it more often than not. Some of the indicated probabilities are not too difficult to prove; others are trickier.

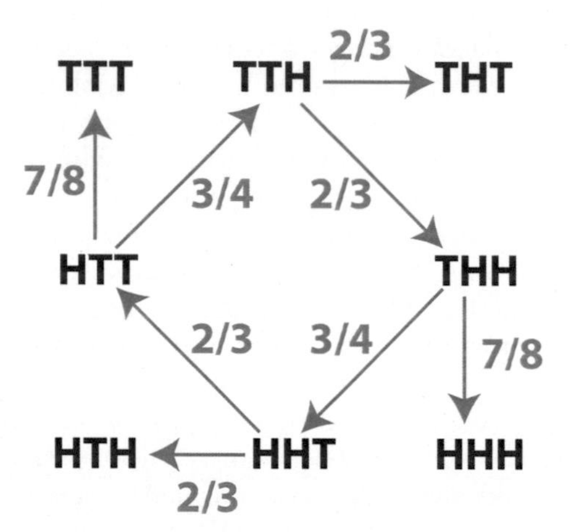

Notice in particular the diamond in the middle of the diagram, representing four sequences in a cycle that better each other. It follows that none of those four sequences can be best overall.

This kind of no-best-choice scenario is referred to as *non-transitivity*, and is actually quite common. It occurs most famously in the game of Rock-Scissors-Paper, where Rock beats Scissors beats Paper beats Rock. Similar Rock-Scissors-Paper competitions can even occur in nature.

Non-transitivity is not in general a surprise, but that it occurs in Penney's game is very counterintuitive. The source of the confusion is considering a long sequence of tosses as a succession of shorter three-toss sequences; those shorter sequences are not independent of each other, and so our intuitive estimate of the probabilities goes awry.

Now, what do visitors get out of Mathamazing's version of Penney's game? Not a whole lot.

The Mathamazing version consists of a barrel containing Red and Blue balls. The visitor punches buttons to choose a three-ball sequence and then the computer chooses its sequence; the balls are then drawn at random until either the visitor's or the computer's sequence appears, and the winner is declared. The computer keeps a running tally of its success rate.

It is hard to see what visitors might get out of simply playing this game once or twice. Sure, the majority of the time the computer wins, but often enough the player wins. Whoever wins, it's hard to see why the response should be anything other than "meh". Yes, Mathamazing's Penney exhibit is accompanied by an explanation. Unfortunately, that explanation is long, poorly worded and confusing.

Although each of the eight possible sequences of three balls IS equally likely, whoever chooses first (Player 1) is more likely lo lose. Why?

Player 2 (here, the computer) knows what Player 1 chose. They can use that information to increase their chance of winnlng.

For example, Player 1 might choose `red, red, red' (RRR). This sequence will occur with the first three balls 1/8 of the time. Player 1 will win and there isn't anything Player 2 can do about it.

However, it only occurs 1/8 of the time, which means it doesn't occur 7/8 of the time. The only way it can not be RRR is if a blue ball gets sampled.

This is how Player 2 can gain advantage by choosing a sequence that starts with 'blue'.

As soon as a blue ball is sampled, Player 1's sequence has to start again and Player 2 will have the first ball of their sequence. Player 2 is ahead, and Player 1 can't catch up.

Similar thinking reveals that no matter what Player 1 chooses, Player 2 can always choose a `better' sequence.

It doesn't mean that Player 1 can't win, they've just got the odds stacked against them.

You can compare these theo-retical values to real scores on the screen that have been

Mathamazing introduces Penney's game with a silly non sequitur question: is it true or false that "In games of chance, if you go first, you're more likely to win"? The reason then provided for the computer's advantage is not exactly wrong, but it's also not quite right. The reasoning implies, for example, that the computer choosing BBB should normally beat RRR, which is false: just by symmetry, the two sequences are equally likely to win.

It hardly matters what is written to explain the game, however, since it is clear that almost no one reads it. Did you, Dear Reader?

Your Maths Master spent a good half hour watching visitors interact with Mathamazing's Penney exhibit. He didn't see a single person, child or adult, do anything other than push the buttons to start the game, watch the machine do its thing, summarize the experience with a "meh" and walk away. Not a single person, child or adult, even attempted to read the provided explanation.

The same was pretty much true of all the Mathamazing exhibits. Your Maths Master really enjoyed many of the exhibits, and he appreciated their cleverness and mathematical merit; if there had been a concerned and cluey parent or teacher or museum guide on hand to explain what was going on, the exhibits could have been great, fun lessons. But that simply didn't seem to be the case. (We've seen guides perform wonders at Melbourne Museum, and we're not entirely sure why they weren't employed at Scienceworks.)

Your Maths Master never witnessed anybody treating any of the displays as more than peculiar pinball machines. No one appeared to be pondering the mathematical significance of the exhibits, and hardly anyone read a word of the explanations. Which turned out to be wise, since almost all of the descriptions were over-wordy and clumsy to the point of uselessness.

We've picked on Scienceworks in other columns,[1] and we somewhat regret doing so again. We appreciate what Mathamazing is attempting something admirable, and some kids probably get something out of the experience. We're just not convinced that many kids get all that much and, with proper and informed guidance, we believe they could get so much more. Moreover, some of the exhibits and their clumsy explanations really need to be rethought; if an exhibit requires so many words to be explained, words that almost no one will read, perhaps the exhibit simply doesn't work. There do exist stunning mathematical exhibits that require few words, or no words at all,[2] and we hope Scienceworks and Questacon are contemplating such exhibits for their future displays.

Your Maths Masters really want to like Mathamazing, and Scienceworks more generally. At the moment, however, we're not convinced that Mathamazing in practice is significantly more educational than a day with the mechanical rides and games at an amusement park. And the amusement park is a lot more fun.

Puzzles to ponder

Show that in Penney's game, HTT has a 3/4 chance of beating TTH. What are the chances of HTT beating THH?

[1]See Chapters 63 and 64 of *A Dingo Ate My Math Book.*

[2]See Chapters 5 and 52.

CHAPTER 41

The Playmobil mystery

A few weeks ago, Junior Maths Master Karl began to collect toy Playmobil®[1] figures. There are twelve figures to collect (in this particular series), and the figures come in identical, opaque packets. This makes each purchase a new surprise. You may be able to guess what happened.

For those old enough to remember the heyday of football cards, the fun and the frustration will be all too familiar. The more cards you purchased, the more difficult it was to obtain the missing few: one more Peter McKenna, and never a sign of that elusive Stan Alves.[2] And, all the while, you would be accumulating a mountain of the ghastliest chewing gum known to mankind.

Back to Karl and Playmobil. Karl recently bought his eighth mystery packet and ... the packets have contained eight different figures. Huh.

That seems distinctly lucky, but exactly how lucky? We've been rounding up all the children who collect Playmobil figures, and we'll be taking a survey to see how they fared. But, while waiting for the children to be herded into the holding pen, let's do a few calculations.

For what follows, we'll make just one assumption: that the makers of Playmobil haven't stacked the deck, that all figures are produced in the same quantities. That means that a given packet is equally likely to contain each of the twelve possible

[1] Playmobil is a registered trademark of Geobra Brandstätter Siftung & Co. KG.

[2] Two star Aussie rules stars of the Victorian Football League from the 1960s.

figures. (We believe this to be true for Playmobil cards, but it is definitely *not* true, for instance, of Pokémon cards.)

Now, it's pretty easy to determine the chances of Karl's eight figures all being different. To begin, there are 12 possible figures contained in the first packet purchased, then again 12 in the second packet, and so on. So, there is a total of $12 \times 12 \times \cdots \times 12$ possible purchasing sequences.

On the other hand, if Karl's figures were all to be different then, whatever the first figure obtained, there are only 11 possible figures for Karl's second purchase, 10 figures for his next purchase, and so on. It follows that the chances of Karl's eight figures all being different is

$$\frac{12 \times 11 \times 10 \times 9 \times 8 \times 7 \times 6 \times 5}{12 \times 12 \times 12 \times 12 \times 12 \times 12 \times 12 \times 12} \approx \frac{1}{22}\,.$$

So, there was a little less than a 5% chance of Karl's eight figures being different. Not very likely, but it should occur for about 1 in 22 kids.

Here's a tougher question: how many packets would we *expect* Karl had to purchase to obtain eight different figures? That's trickier, so we'll begin with a warm-up. Suppose we have a die, and we're going to keep rolling it until a 1 appears. On average, how many times will we have to roll the die?

Each roll of the die gives a 1/6 chance of coming up 1. So, perhaps it should take on average 6 rolls to obtain our 1? Well, it's actually more complicated than that. In a rare instance, however, of probability not leading us astray, the answer 6 is actually correct. This is not too difficult to justify in a careful manner, but we'll leave that for another occasion.[3]

With the above in mind, it is now straight-forward to calculate Karl's expected Playmobil purchases. His first purchase is obviously fine, no matter which figure he obtains. Then, in the next packet Karl is hoping for any of the other eleven figures, and the chances are 11/12 of that occurring. So, following our dice example, it will take on average 12/11 purchases to obtain a second figure.

Having obtained two distinct figures, the chances of obtaining a third figure are then 10/12 with each packet. So, on average it will take 12/10 extra packets to go from having two distinct figures to three. Continuing in this manner, we see that to have obtained eight distinct figures, the average number of packets Karl would need to have purchased is

$$1 + \frac{12}{11} + \frac{12}{10} + \frac{12}{9} + \frac{12}{8} + \frac{12}{7} + \frac{12}{6} + \frac{12}{5} \approx 12\,.$$

So, although it was pretty lucky for Karl to have needed only eight packets to obtained eight different figures, we wouldn't expect him to have required many more than eight packets to do so.

But of course from here on it gets much harder. To collect all twelve figures, the average number of packets required is

$$1 + \frac{12}{11} + \frac{12}{10} + \cdots + \frac{12}{2} + \frac{12}{1} \approx 37\,.$$

[3]See the next Chapter.

Now, Karl has a great start. Even so, to obtain those last four figures, the number of extra packages that Karl (and his very patient father) can expect to buy is

$$\frac{12}{4}+\frac{12}{3}+\frac{12}{2}+\frac{12}{1}\approx 25\,.$$

Well, Karl (and his very patient father) have already begun. And, Karl's ninth packet contained ... a ninth distinct figure. Hmm.

It turns out that Junior Maths Master Karl is very cunning. Rather than trust in the vagaries of luck, Karl simply felt each packet, to determine the shape of the figure enclosed. Sometimes probability is so beautiful, and so much not the point.

Puzzle to ponder

In 1966, the last year that Aussie rules football really mattered,[4] there were 72 playing cards to collect, and they were purchased three cards, plus gum, to a pack. On average, and not counting trading with friends, about how many packets would you expect to need to purchase to obtain a complete set? What would be the total nutritional content of the chewing gum obtained in the process?

[4]That was the last time, and only time, St. Kilda won the Premiership.

CHAPTER 42

The devil is in the dice

Undoubtedly the most annoying children's game of all time is Ludo. Yes, it's fun once you get going, but you can't get going until you roll a 6. And how long does it take on average to get that 6? The official answer from our youthful sources is "forever".

We pondered this question recently,[1] and our answer then was a bit more optimistic. Given that we have a 1/6 chance with each roll of the die, a reasonable guess is that it will take on average six rolls to get our 6. Now, proving that guess correct provides us with the opportunity to demonstrate some very clever mathematics.

Let's consider the possibilities step by step. To begin, the chances of needing just one roll, by getting a 1 on the first shot, are just 1/6.

What about the chances of needing exactly two rolls? That would involve getting anything other than a 1 on the first roll – there's a 5/6 chance of that – and then succeeding on the second roll. So the chances of needing exactly two rolls are $5/6 \times 1/6 = 5/36$.

Similarly, the chances of needing exactly three rolls – a miss then another miss then a hit – are $5/6 \times 5/6 \times 1/6 = 25/216$. Then the chances of needing exactly four rolls are $(5/6)^3 \times (1/6) = 125/1296$, and so on.

To find the average number of rolls we have to sum up all the chances for the different numbers of rolls. So, the average is given by the infinite sum

$$\frac{1}{6} + \left(2 \times \frac{5}{6} \times \frac{1}{6}\right) + \left(3 \times \left(\frac{5}{6}\right)^2 \times \frac{1}{6}\right) + \left(4 \times \left(\frac{5}{6}\right)^3 \times \frac{1}{6}\right) + \cdots .$$

[1] See the previous Chapter.

That looks like a ton of work to add up. However, here's where we employ a very clever trick.

Let's consider where we might be after the first roll. If we've rolled a 6 then we're done with just the 1 roll. But there's a 5/6 chance of missing and we're back where we started. So, if we let A stand for the average number of rolls required, then A must satisfy the equation

$$A = 1 + \left(\frac{5}{6} \times A\right).$$

Multiplying both sides by 6, we easily solve to give $A = 6$.

That's a lovely proof but did we really need to bother? Why not just be satisfied with our intuition? Let's consider a related question: suppose we want to roll a 6 twice in a row? On average how many rolls would be required for that?

The chances of getting 6 on any two rolls is $1/6 \times 1/6 = 1/36$. So, if we take 36 double-rolls, 72 rolls in total, should that be what we need on average to obtain a double-6? Nope, that answer is wrong.

The answer, to life, the universe and everything, *and* the number of dice rolls, is 42. So much for intuition.[2]

Such examples are discussed by mathematician Theodore Hill in a beautiful article on faking data.[3] Most people think of random data such as dice rolls and coin flips as the very opposite of streaks, with lots of alternating outcomes: heads-tails-heads, and so forth. In fact the opposite is true, and randomness is very streaky. And this failure of intuition makes it very difficult for people to fake data.

To illustrate, here is a wonderful demonstration employed by Hill. Ask your students, friends, family members, pets, anyone to flip a coin 100 times and record the sequence of outcomes. Alternatively, they can just pretend to do the exercise and write down the "results" of 100 fake tosses. Then, simply by looking at the results, you can very reliably determine whether they chose to flip for real or to fake it.

When flipping a real coin, the chances that 100 flips will contain a streak of five heads is about 81%. And, the chances that there is either a sequence of five heads *or* five tails (or both) is about 97%. By contrast, almost always when people attempt to fake data they avoid streaks, long sequences of heads or tails. So, the more alternating data is very likely to be fake, and the streaky data is probably real.

We human beings are just very poor probability machines.

Puzzles to ponder

Would you guess that the tail of dice pictured at the beginning of this chapter is random or faked?

Can you prove our claim above, that on average it takes 42 rolls of a die to get 6 twice in row?

---

[2]But Douglas Adams must be up there, somewhere, smiling.

[3]*The Difficulty of Faking Data*, Chance, **26**, 8–13, 1999.

CHAPTER 43

The Freddo Frog path to perfection

First semester was very stressful, and not just for the uni students. One of your Maths Masters took over the lecturing and administration of a major first year subject, which meant writing new assignments, new tutorial sheets and new exams. And, to save students the cost of a textbook, your Maths Master also decided to produce 300 pages of new lecture notes. (Your other Maths Master considers this decision to have been barking mad.)

Of course having hundreds of rapidly written pages guarantees there will be plenty of glitches and (hopefully minor) mathematical goofs. To help find them your cunning Maths Master enlisted his hundreds of students as proofreaders: every error spotted was rewarded with a Freddo Frog.[1]

Such reward schemes have been employed by many academics, perhaps most bravely by Stanford computer scientist Donald Knuth. Many readers have probably never heard of Knuth but he is a hero to mathematicians, almost all of whom use TeX, Knuth's brilliant typesetting software.[2] Think of Knuth as like Bill Gates, except that Knuth's products don't suck, they're free, and Knuth is not busy stuffing up American education.

Knuth has offered various reward schemes for error-spotting. For his TEX program, Knuth initially offered a reward of $1.28 for each error, which was then doubled each year to a maximum of $327.68 per error. This is reminiscent of the famous wheat and chessboard problem.[3] True, Knuth capped the doubling, so it was unlikely to send him broke, but he has still issued tens of thousands of dollars worth of checks. Luckily for Knuth, most of the recipients have kept his checks as treasured mementos, rather than cash them.

Now, suppose you're just finishing up your masterpiece and you want to employ such a reward scheme. It would be reassuring to have some idea of the number of

---

[1] A very old and popular Australian chocolate bar, now made by Cadbury. It's not very good.

[2] This book was typeset with TeX.

[3] Place one grain of wheat on the first square of the board, 2 on the second square, 4 on the third, and $\cdots$ reap the harvest on the 64th square.

errors for which you'll be paying. It turns out this type of problem is very common and can be handled in a very elegant manner.

Let's consider a different problem. Imagine you're an ecologist and you want to count the number of fish in Lake Woebegon. Of course you could just drain the lake and pile up the dead fish, however your fellow ecologists would probably frown upon this method. So, what you can do instead is catch a bunch of the fish, tag them and release them. Let's imagine, for example, that you tag and release 900 fish.

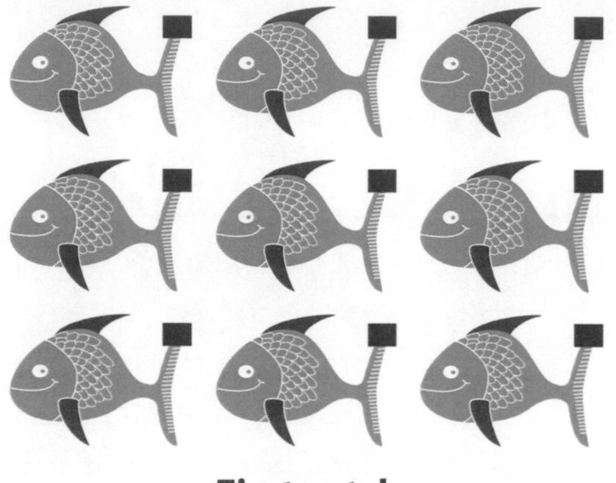

First catch

Next, after allowing sufficient time for the tagged fish to become reacquainted with their friends, you make a second haul of fish. Let's suppose this time you catch 1500 fish, of which 300 are tagged.

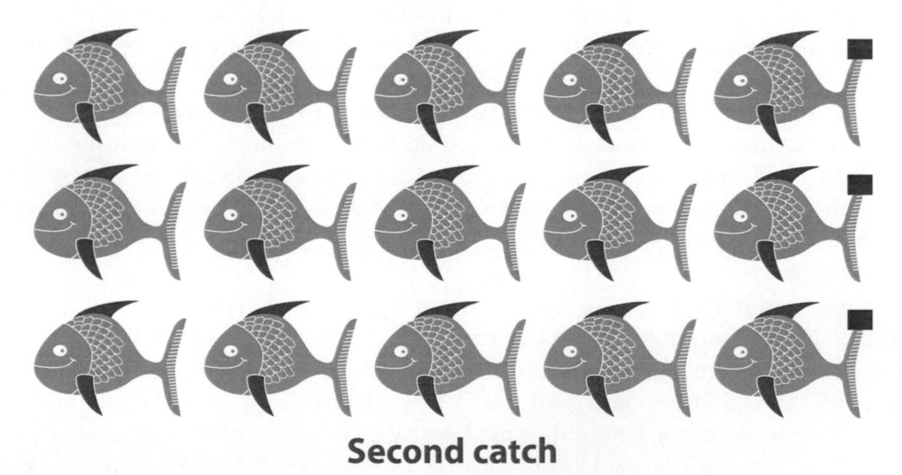

Second catch

So, you ended up catching 1/3 of the tagged fish that you had released. This suggests that the 1500 fish in your second haul should be about 1/3 of the total number of fish in the lake, making for about 4500 fish. Problem solved, roughly, and no dead fish.

Now, back to your masterpiece. You can use the above tagging method to estimate the number of errors by employing two initial proofreaders, Bert and Ernie. Bert effectively tags some errors for Ernie to catch, and vice versa.

To make it concrete let's suppose Bert finds 20 errors and Ernie finds 30 errors. Suppose further that there are 5 common errors, included in both their lists. That means Ernie found 1/4 of Bert's errors, and so we estimate that Ernie's 30 errors account for about 1/4 of the errors in the book. That suggests a total of about 120 errors.

Forms of this tag-and-release strategy have been employed since the 19th century and are now very commonly used when estimating population numbers. This has included obviously important work, such as estimating the prevalence of childhood diabetes in a city. It has also been used in fascinating though less important studies, such as estimating the size of Shakespeare's vocabulary. For all such applications there have been developed sophisticated statistical methods to account for changing populations, repeated tagging, biased tagging and so forth. You might ponder, for example, if Bert and Ernie are likely to find and to skip over the same sort of errors, and how one might adjust for that.

Of course all this does is estimate a population size, and if one's masterpiece is to be perfect then each and every error must still be caught. Which can take some time: Sir Isaac Newton's *Principia Mathematica*, perhaps the most important mathematics book ever written, contained an incorrect equation that went unnoticed for three hundred years.[4] If only Sir Isaac had been more generous with his Freddo Frogs.

Puzzle to ponder

In the masterpiece example, we imagined errors tagged by Bert, and used Ernie's "capturing" of them to estimate the total errors. Show that we obtain the same total estimate if we use Bert to capture Ernie's tagged errors. More generally, suppose Bernie find B errors, Ernie finds E errors, and that C errors are common to both their lists. Then show that the two ways of estimating the total number of errors result in the same answer.

[4]R. Garisto, *An Error in Isaac Newton's Determination of Planetary Properties*, American Journal of Physics, **59**, 42–48, 1991.

CHAPTER 44

Will Rogers, clever Kiwis and medical magic

Sir Robert Muldoon was a memorable, pugnacious prime minister of New Zealand. In a typically diplomatic moment, "Piggy" Muldoon commented on his many emigrating countrymen:

"New Zealanders who leave for Australia raise the IQ of both countries."

Is that possible? Well, a New Zealander with IQ lower than New Zealand's average would raise that average by leaving. And, if this same New Zealander had an IQ higher than Australia's average, their arrival would also raise Australia's average.

So, Muldoon's is a very clever double-joke, simultaneously having a go at both Australians and emigrating New Zealanders. Unsurprisingly, Muldoon's joke has been used in many other contexts, and unsurprisingly it's not really Muldoon's joke.

It seems that the joke originated with the great American humorist Will Rogers. Born in Oklahoma, Rogers was describing his fellow "Okies" emigrating to California during the Dust Bowl of the 1930s.

However, our simple analysis above demonstrates that there is also a serious principle underlying Rogers' excellent joke. Given two collections of objects, it may be possible to raise – or lower – the averages of both collections simply by switching some of the objects from one collection to the other.

This principle, now known as the *Will Rogers phenomenon*, has some important and surprising applications. For example, in the diagnosis of diseases such as cancer, patients are often categorized into different "stages". Then, advances such as improved diagnostic techniques can alter the manner in which patients are categorized, resulting in *stage migration*. In turn, this stage migration can lead to an illusory improvement in recovery rates.

To illustrate, suppose we have a number of patients suffering a particular illness. Imagine that a third of our patients are very ill and even with prolonged treatment only 30% will fully recover. Another third of the patients are only slightly ill, and 70% of them would make a full recovery with little or no treatment. Finally, there is a middle, moderately ill group, 50% of whom would recover with or without treatment.

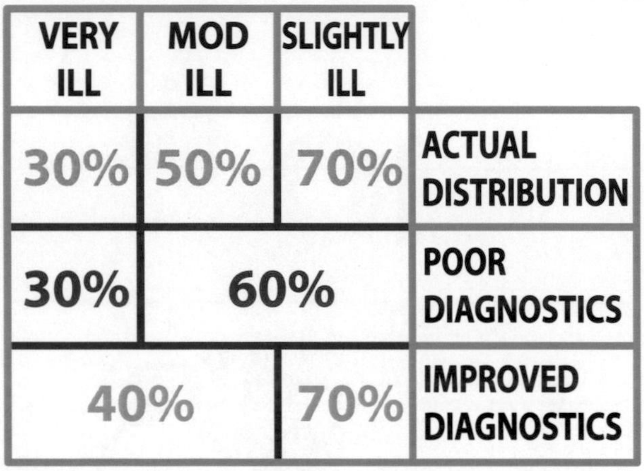

<table>
<tr><th>VERY ILL</th><th>MOD ILL</th><th>SLIGHTLY ILL</th><th></th></tr>
<tr><td>30%</td><td>50%</td><td>70%</td><td>ACTUAL DISTRIBUTION</td></tr>
<tr><td>30%</td><td colspan="2">60%</td><td>POOR DIAGNOSTICS</td></tr>
<tr><td colspan="2">40%</td><td>70%</td><td>IMPROVED DIAGNOSTICS</td></tr>
</table>

Now, suppose our diagnostic techniques are poor, and that we cannot distinguish the slightly ill and moderately ill patients. In that case, we'd put them altogether in one group, with a combined recovery rate of 60%.

But then our diagnostics improve, and we can now determine the symptoms of the moderately ill patients. We decide these patients should receive the same treatment as the seriously ill patients. The recovery rate of those "serious" patients will then rise from 30% to 40%, and simultaneously the recovery rate of the "slightly ill" patients will have risen from 60% to 70%. Magic!

Now, if you're a doctor witnessing your patients at all stages doing better on average, you may innocently but erroneously attribute this "improvement" to your brilliant doctoring. Much less innocently, the manager of a hospital might "improve" the performance of the hospital's intensive care unit, and of the rest of the hospital, by simply weakening the admission criteria for the intensive care unit. The opportunities for error and abuse are endless.

The Will Rogers phenomenon is just one more example of how averages and other statistical quantities can so easily confuse and mislead. We've written about such trickery time and time again.[1] So, maybe we've written enough, and perhaps it's time to stop?

Last week, Prime Minister Gillard delivered a very important speech, which began:

We are a nation of 23 million people, whose median age is 37, an average which continues to rise.

No, it's probably too soon to stop.

[1]See Chapter 40, and Parts 5 and 6 of *A Dingo Ate My Math Book*.

Puzzle to ponder

Suppose that the 50 students in Group A average 30% on a test, and the 50 students in Group B average 70%. By moving some students from Group B to Group A, how high might you be able to make the new averages?

CHAPTER 45

The hidden karma of Snakes and Ladders

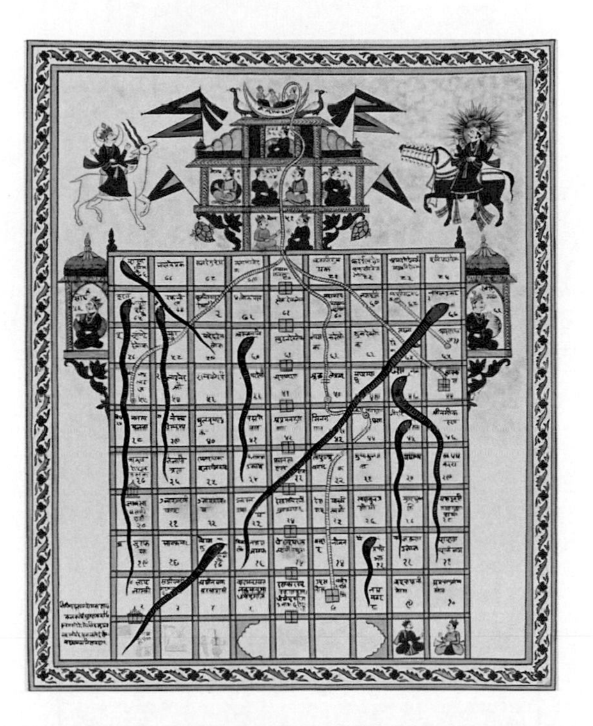

In a previous column your Maths Masters claimed to enjoy the occasional game of Snakes and Ladders.[1] We lied. Though our little maths masters may delight in the vagaries of pure chance, most grown ups, including your Maths Masters, will only play grudgingly, all the while pleading "Dear God, when will this end?"

It turns out that a god is just the right sort of being to answer that question. Snakes and Ladders derives from the old Indian game *gyan chaupar*, which was imbued with Hindu (or Jain or Muslim) symbolism. The ladders, labelled "knowledge" and "devotion" and so forth, led one to Vaikuntha, the home of the god Vishnu; in opposition, the snakes of "darkness" and "illusion" and the like led the player further away. (In a similar manner many nineteenth century English versions of the game were cloaked in a Victorian morality.)

Alas, the gods never bothered to tell us how long a game actually takes. But luckily, mathematicians have come to the rescue.[2] To indicate how it works we'll investigate your Maths Masters' version, the game of (singular) *Snake and Ladder*.

---

[1] Chapter 57 of *A Dingo Ate My Math Book*.

[2] S. C. Althoen, L. King and K. Schilling, *How Long Is a Game of Snakes and Ladders?*, Mathematical Gazette, **77**, 71–76, 1993.

In contrast to the familiar 10×10 setting, Snake and Ladder is played on a pleasingly small 3×3 board, and features just one snake and one ladder:

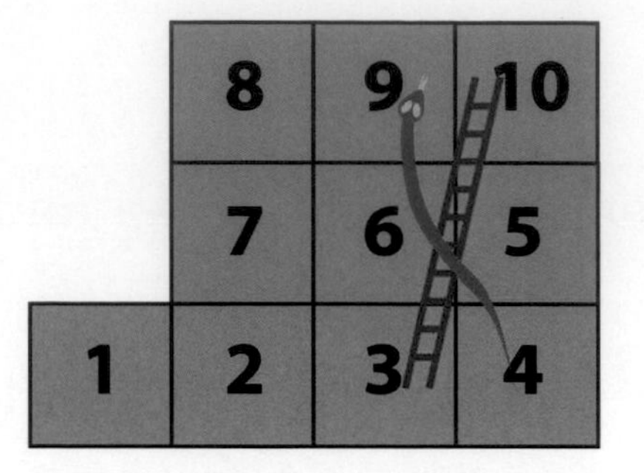

Including the starting 1 square, there are ten squares in total, and as usual rolling a die indicates how many squares to move. We'll treat Snake and Ladder as a solo game, which it effectively always is, since one person's progress does not affect any others'.

Let's first ignore the snake and the ladder. Of course each number on the die has a $1/6$ chance of being rolled, and then the probabilities of moving from here to there on a given turn are captured by the following 10×10 table:

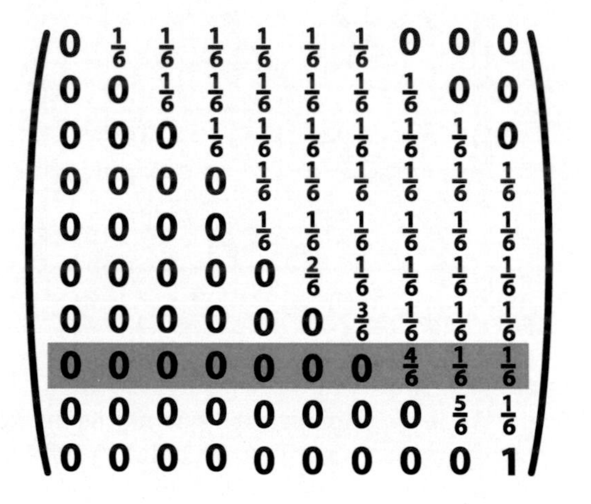

$$\begin{pmatrix} 0 & \frac{1}{6} & \frac{1}{6} & \frac{1}{6} & \frac{1}{6} & \frac{1}{6} & \frac{1}{6} & 0 & 0 & 0 \\ 0 & 0 & \frac{1}{6} & \frac{1}{6} & \frac{1}{6} & \frac{1}{6} & \frac{1}{6} & \frac{1}{6} & 0 & 0 \\ 0 & 0 & 0 & \frac{1}{6} & \frac{1}{6} & \frac{1}{6} & \frac{1}{6} & \frac{1}{6} & \frac{1}{6} & 0 \\ 0 & 0 & 0 & 0 & \frac{1}{6} & \frac{1}{6} & \frac{1}{6} & \frac{1}{6} & \frac{1}{6} & \frac{1}{6} \\ 0 & 0 & 0 & 0 & \frac{1}{6} & \frac{1}{6} & \frac{1}{6} & \frac{1}{6} & \frac{1}{6} & \frac{1}{6} \\ 0 & 0 & 0 & 0 & 0 & \frac{2}{6} & \frac{1}{6} & \frac{1}{6} & \frac{1}{6} & \frac{1}{6} \\ 0 & 0 & 0 & 0 & 0 & 0 & \frac{3}{6} & \frac{1}{6} & \frac{1}{6} & \frac{1}{6} \\ 0 & 0 & 0 & 0 & 0 & 0 & 0 & \frac{4}{6} & \frac{1}{6} & \frac{1}{6} \\ 0 & 0 & 0 & 0 & 0 & 0 & 0 & 0 & \frac{5}{6} & \frac{1}{6} \\ 0 & 0 & 0 & 0 & 0 & 0 & 0 & 0 & 0 & 1 \end{pmatrix}$$

As an example of how to read the table, if we're on the 8th square then the chances of where we'll end up are indicated by the 8th row. Of course with no snakes around there is no chance that we'll wind up on the same square or any earlier square, which is indicated by the zeroes at the beginning of the row. We also have a $1/6$ chance of throwing a 1 to take us to square 9, and there's a $1/6$ chance of rolling a 2, landing us (finally!) on square 10. There's also a $4/6 = 2/3$ chance that we'll throw a 3 or higher, overshooting the final square and meaning we have to stay put. (The finicky S & L gods have decreed that in order to finish the game a player must land exactly on the final square.)

Let's now adjust the table to take account of the snake and the ladder. If we ever land on square 3 the ladder immediately teleports us to square 10 (yay!). So, it is impossible to wind up on square 3 at the end of a turn, and we have a correspondingly increased chance of arriving at square 10. To account for this we add the 3rd column of the table to the 10th column, and then the 3rd column is converted to all zeroes. In the same manner the snake is handled by adding the 9th column to the 4th column and then zeroing the 9th column. These adjustments result in the following 10×10 table:

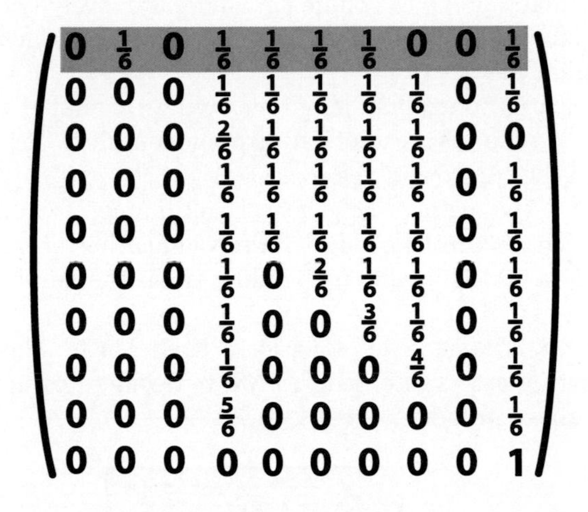

$$\begin{pmatrix}
0 & \frac{1}{6} & 0 & \frac{1}{6} & \frac{1}{6} & \frac{1}{6} & \frac{1}{6} & 0 & 0 & \frac{1}{6} \\
0 & 0 & 0 & \frac{1}{6} & \frac{1}{6} & \frac{1}{6} & \frac{1}{6} & \frac{1}{6} & 0 & \frac{1}{6} \\
0 & 0 & 0 & \frac{2}{6} & \frac{1}{6} & \frac{1}{6} & \frac{1}{6} & \frac{1}{6} & 0 & 0 \\
0 & 0 & 0 & \frac{1}{6} & \frac{1}{6} & \frac{1}{6} & \frac{1}{6} & \frac{1}{6} & 0 & \frac{1}{6} \\
0 & 0 & 0 & \frac{1}{6} & \frac{1}{6} & \frac{1}{6} & \frac{1}{6} & \frac{1}{6} & 0 & \frac{1}{6} \\
0 & 0 & 0 & \frac{1}{6} & 0 & \frac{2}{6} & \frac{1}{6} & \frac{1}{6} & 0 & \frac{1}{6} \\
0 & 0 & 0 & \frac{1}{6} & 0 & 0 & \frac{3}{6} & \frac{1}{6} & 0 & \frac{1}{6} \\
0 & 0 & 0 & \frac{1}{6} & 0 & 0 & 0 & \frac{4}{6} & 0 & \frac{1}{6} \\
0 & 0 & 0 & \frac{5}{6} & 0 & 0 & 0 & 0 & 0 & \frac{1}{6} \\
0 & 0 & 0 & 0 & 0 & 0 & 0 & 0 & 0 & 1
\end{pmatrix}$$

The above table, which we'll call T, is referred to as a *transition matrix*, and Snake(s) and Ladder(s) is what is known as a *Markov process*. (It's a genuine Markov process, as opposed to the ill-conceived scenarios one tends to find on Victorian exams.) The point is that the above matrix tells us everything we need to know about the probabilities of how the game will progress.

As an example, if we want to know the probability of finishing the game within the first two moves we can calculate $T \times T = T^2$, the square of the matrix T:

$$\begin{pmatrix}
0 & 0 & 0 & \frac{5}{36} & \frac{1}{12} & \frac{5}{36} & \frac{7}{36} & \frac{5}{36} & 0 & \frac{11}{36} \\
0 & 0 & 0 & \frac{5}{36} & \frac{1}{18} & \frac{1}{9} & \frac{1}{6} & \frac{2}{9} & 0 & \frac{11}{36}
\end{pmatrix}$$

The top part of T^2 is pictured above, and the final number in the first row indicates that there is a $11/36$ chance of finishing the game on the first or second move. Similarly, higher powers of T tell us the chances of finishing the game within three moves, or four moves, and so on. (Since the 3rd and 9th rows and columns of T are redundant, in practice we'd delete them and work with the resulting 8×8 matrix.)

But how long on average will a game last? There is no theoretical limit to the length of a game, which suggests that we need to calculate all possible powers of T and somehow sum the results. That sounds tricky and tedious. Sometimes, however, infinite sums have very finite, easily obtainable answers. Such is the case here.

Snakes and Ladders is what is known as an *absorbing Markov process*: once we arrive at square 10, we simply stay there. That turns out to make the probability calculations very pretty. The (very not obvious) trick is to lop off the last row and column of T, creating a "reduced" 9×9 matrix R. Our problem is then to calculate the sum of all the powers of R:

$$1 + R + R^2 + R^3 + \cdots.$$

The beginning 1 of this sum stands for the so-called *identity matrix*, which in the world of matrices behaves very much like our everyday number 1: if M is any other matrix with the same dimensions, then $1 \times M = M$, and so on.

The above is a *geometric sum* of matrices. We have encountered geometric sums of numbers in many previous columns.[3] There are lovely tricks for calculating these sums and similar tricks can be applied to calculate geometric matrix sums, producing very similar answers.

It's a bit complicated to explain here, but having calculated the matrix output of this infinite sum, we can just read off the average length of our game. We simply add the numbers in the top row of the matrix.

It turns out that the average length of a solo game of Snake and Ladder is exactly six moves. That's pretty good, definitely much better than the average of around 40 moves for a 10×10 game (depending upon the number and location of the snakes and ladders).

Still, even our 3×3 game is too long for a tired Maths Master, just wanting to get the little maths masters into bed. So, we've decided to shorten our game by sneakily including an extra ladder, from 2 to 7.

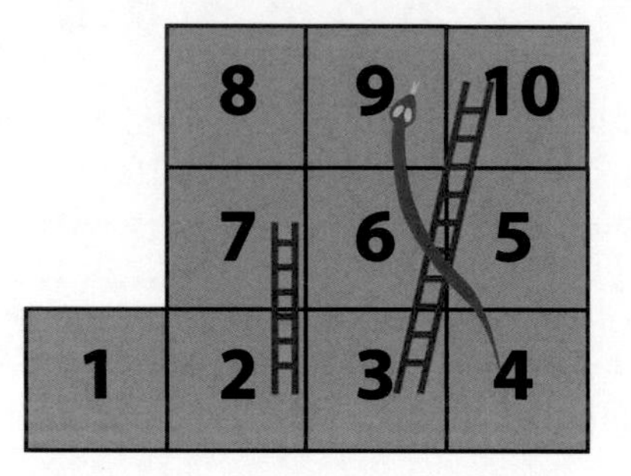

It is easy to adjust the transition matrix to account for our cunning little ladder. Then, summing the matrix powers as before, we find that the average length of our new game is ... exactly six moves.

Huh? How could our new ladder have had no effect?

In fact, the ladder has had two effects. First, as intended, the new ladder gave us a chance of leaping closer to the finish. However a second effect was to create a chance of bypassing the long ladder from square 3 to the finish. As it happened, these two effects exactly cancelled. On larger boards this second effect can easily be much more pronounced, resulting in a significant lengthening of the average game, and thwarting all attempts to hustle the little maths masters off to bed. Similarly, a sneaky snake can actually shorten the average game.

The gods of Snakes and Ladders work in wondrous ways.

Puzzle to ponder

Our T^2 matrix indicates that there is a $11/36$ chance of finishing a game of Snake and Ladder in either one or two moves. Show that this is true.

[3] See, for example, Chapter 24.

Part 6

Space and Time, Final Frontiers

CHAPTER 46

Poet of the Universe

Welcome back, for the sixth year of Maths Masters.[1] We're very grateful for the continued interest that our columns have received.

Again this year, we hope to write about some beautiful and little-known mathematics, and to explain more familiar math that tends to confuse. And yes, we'll probably wind up whacking a windmill or two. However, we've chosen to begin the year by honoring a very special mathematician.

In the 1980s, one of your maths masters was lucky enough to be given the opportunity to study at Stanford University. Soon after arriving, he met a mathematician who looked remarkably like Groucho Marx, complete with stooped walk, and a nasal, jokey manner. Robert Osserman turned out to be almost nothing like Groucho. Bob definitely had a great sense of humor, but he was also gentle and generous. As it happened, your maths master began researching *minimal surfaces* – a mathematical idealization of soap films – a field in which Bob was an expert. Your maths master found himself asking Bob question after question. He read Osserman's excellent (technical) introduction to minimal surfaces, many times over.[2]

None of this is exceptional, merely indicative of an accomplished scholar and dedicated teacher. But then an opportunity arose.

As do most American universities, Stanford encourages undergrads to at least have some breadth in their studies. So, science students must read and write upon

[1]This column was written in 2012.

[2]*A Survey of Minimal Surfaces*, Dover, 2014.

at least a few great books, and arts students are required to study a little math and science.

University subjects are typically ill-suited to outsider students, so it is common for American universities to offer specialized "breadth subjects". Bob Osserman, together with Sandy Fetter (physics) and Jim Adams (sociology) introduced such a year-long subject at Stanford: *The Nature of Mathematics, Science and Technology.* Your maths master was lucky enough to be one of the two tutors.

It was a wonderful subject, with the three lecturers perfectly complementing each other: Bob introduced the abstract mathematical ideas; Sandy demonstrated how these ideas could be used to model the physical world; and Jim described the often circuitous path from the physics to technological breakthrough. This was gently and engagingly presented to a hundred nervous arts students, most of whom were ready to flee at the first appearance of an equation.

Your maths master is undoubtedly biased, but he found Bob Osserman's presentations particularly enjoyable and insightful. Bob presented mathematics as a collection of ideas, and as the history of the mathematicians who struggled, not always successfully, to untangle those ideas. Bob used the arts, and music in particular, to motivate the mathematics. He began with the Pythagorean theory of harmony, based upon simple ratios of string lengths: $2/1$ and $3/2$, and so on. Bob then demonstrated how that simple idea leads to more complicated ratios, and finally, almost in contradiction, to irrational numbers. This fascinating history also included a guest appearance by Vincenzo Galilei, the father of his much more famous son.

Bob, Sandy and Jim had planned to write a textbook for their subject but unfortunately it never eventuated. Bob, however, did write *Poetry of the Universe*,[3] a wonderful book intended for the general public.

Poetry of the Universe tells the history of the mathematical exploration of the Universe. Osserman begins the story in 240 BC, with the Alexandrian mathematician Eratosthenes and his stunningly simple estimation of the size of the Earth.

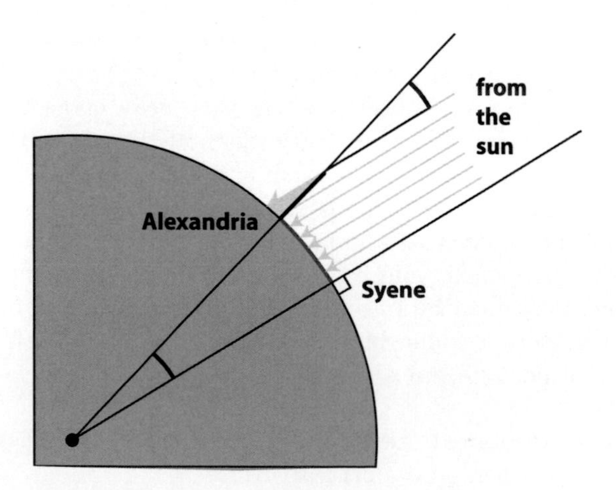

Eratosthenes knew that almost due South of Alexandria was the city of Syene, with a notable feature: at noon on the summer solstice – the day of the year

[3]Anchor, 1996.

when the North pole is tilted most towards the Sun – the Sun is directly overhead. Eratosthenes then measured that in Alexandria at that same time, the Sun makes an angle with the vertical of a little over 7 degrees, about 1/50 the way around a circle. From this, Eratosthenes deduced that the circumference of the Earth must be about 50 times the distance from Syene to Alexandria. That distance, in modern units, is about 800 kilometers, leading to a remarkably accurate estimate of the Earth's circumference of about 40,000 kilometers.

Poetry of the Universe takes the story right up to the present day exploration of the shape of the Universe. Along the way there is wonderful history and fascinating digressions, perhaps the most fascinating of which is the story of Dante's universe.

In *Paradiso*, the 14th Century poet Dante described the universe as consisting of two spheres. The first sphere is our physical world, with the Earth at its center, and the outer layer of this sphere is the *primum mobile*. The second, heavenly sphere is the *Empyrean*, inhabited by the angels and with a godly point of light at its center.

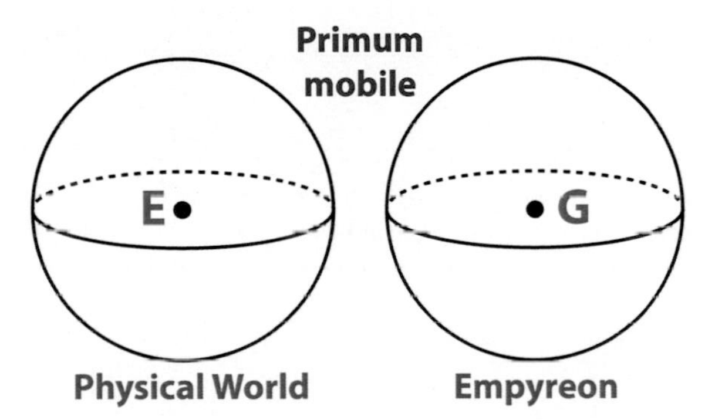

What is fascinating is that the primum mobile, the outer layer of the physical world, is simultaneously the outer layer of the Empyrean. This may be very difficult to imagine, but modern mathematicians know Dante's universe very well.

As Osserman explains, Dante has perfectly described his universe as a *hypersphere*. Analogous to the borderless, 2-dimensional surface of the Earth, a hypersphere is a borderless 3-dimensional world. It is possible that our Universe is a hypersphere, and the hypersphere is the central character of the famous *Poincaré Conjecture*.[4]

In later years, Osserman was Special Projects Director of MSRI one of the World's most prestigious and successful mathematical institutes. In this role, Bob became much more involved in presenting mathematics to the wider community. He engaged in public conversation with such disparate artists as actor Alan Alda and pianist Christopher Taylor, comedian Steve Martin, composer Philip Glass and playwright Tom Stoppard. Bob Osserman recognized mathematics as an art, and he helped many others to see it that way.

Robert Osserman died at his home last November, at the age of 84. Your maths masters will spend the year trying to live up to the brilliant, beautiful example set by this wonderful man.

[4]A famous and notoriously difficult mathematics problem, solved in 2006 by the Russian mathematician, Grigori Perelman.

Puzzle to ponder

If you tried to split the Earth's surface in two, a "physical world" and an "Empyrean", what would these two worlds look like, and what would the "primum mobile" dividing these worlds look like?

CHAPTER 47

Escape to our Moon planet

Your Maths Masters have decided to leave Earth. Yep, it's true. Disillusioned with how our patch of Earth is being run, specifically a small patch in Canberra,[1] we're keen to find a home on a new planet. So, your Maths Masters are headed for the Moon.

And now you begin to doubt us, right? Sure, rocketing thousands of miles into space to escape a nasty administration and a woeful mathematics curriculum makes perfect sense. But escape to our *planet* Moon? Since when has the Earth's moon been a planet?

Well, what makes a planet a planet, or a moon a moon? As illustrated by the heated battle over Pluto's planetary status,[2] these notions are imprecise and contentious. The basic idea, however, is familiar: a planet is a big roundish thing that orbits the Sun, and a moon is a big, but not too big, roundish thing that orbits a planet.

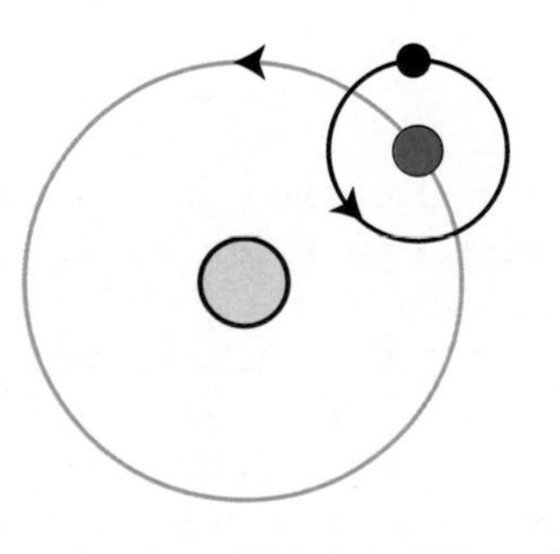

[1] The capital of Australia, and the location of the Federal government.

[2] See Chapter 50.

The path of a moon around the Sun is then the combination of the two simpler orbits. That suggests that, at times, the moon might be traveling "backwards", resulting in a loopy Spirograph path, as pictured below.

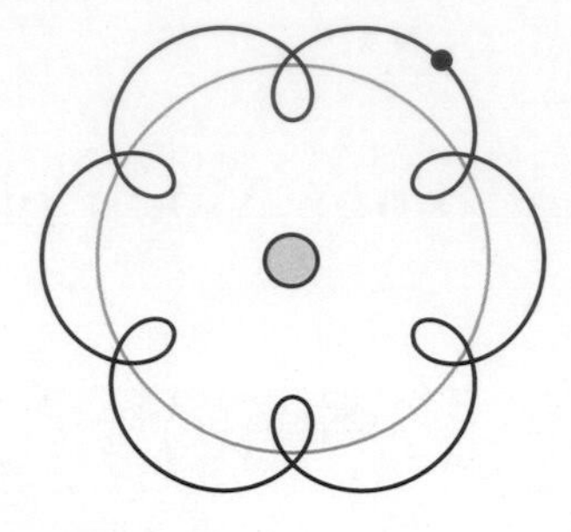

A number of moons in our Solar System follow (roughly) such a path. However not all do, and in particular the path of Earth's moon is notably different from those of all its cousins.

A moon's path around the Sun depends upon a number of factors: the radius of the planet's orbit around the Sun, and the radius of the moon's orbit around its planet; the length of the planet's "year" (the time to travel around the Sun) in relation to the moon's "month" (the time to travel around the planet); and how close the moon and planet are to orbiting in the same plane. It follows that there are *many* possible types of moon orbit.

To get a sense of it all, let's return to our original picture, with the moon and planet orbiting along circles in the same plane. This is approximately true for most, but not all, of the Solar System's moons. Now let's consider what happens if we tinker with the radius of the moon's orbit.

If the radius is large then the moon indeed follows a loopy Spirograph path (although the beginning and end points may not match). As the radius gets smaller so do the loops. For just the right radius the loops vanish completely, producing a cusped path as pictured on the left below.

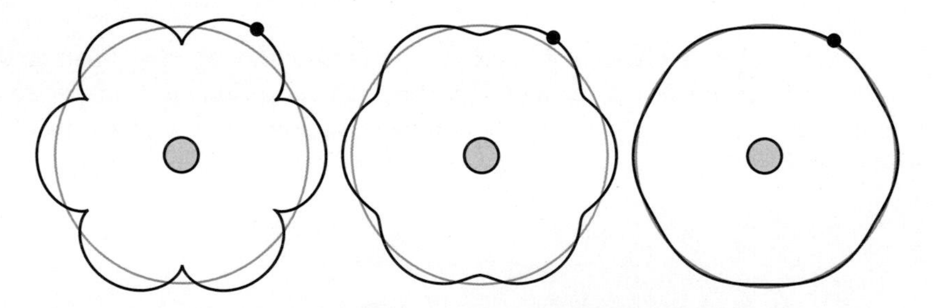

Making the radius even smaller, the cusps turn into smooth wiggles. Then, for even smaller radii, the wiggles pretty much disappear, resulting in a "convex" orbit, without indentations, similar to that of a planet.

So we have four basic types of moon path, and all four occur in our Solar System.[3] Saturn's Mimas exhibits loops, Neptune's Triton follows a very cuspy path, and Jupiter's Callisto has a wiggly path. And, the path of the Earth's moon is of the fourth, convex type; our Moon's path is essentially a rounded dodecagon,

[3]Thanks go to our friend and colleague Don Koks, for assisting our ponderings here.

the twelve sides corresponding to the (approximately) twelve Lunar months in an Earth year. Moreover, the Earth's moon is the only moon in the Solar System that follows a convex path.

The Moon's unique planetlike orbit is a good reason for regarding the Moon as a planet. Moreover, the underlying cause of that orbit provides a second reason; the Sun's gravitational pull on the Moon is actually twice as strong as that of the Earth. Again, the Moon is unique in this regard, with all other moons in the Solar System dominantly controlled by their associated planet rather than the Sun.

That's all very clever, but of course it doesn't change the basic facts: as (almost) every schoolchild knows, the Moon orbits around the Earth, which in turn orbits around the Sun. Well, it turns out the basic facts are not quite that basic.

It is not actually the Earth that orbits the Sun, but more precisely the *center of mass* of the Earth-Moon system that does so.[4] This center of mass is located about 4,600 km from the Earth's center, a good distance towards the Earth's surface. Both the Earth and the Moon travel along paths that oscillate around the orbit of this center of mass. It means, unlike all other moon-planet systems, the Earth's and the Moon's paths around the Sun are largely indistinguishable.

So, why is our Moon not considered to be a planet? Well, indeed it was until 1543, when the great Nicolaus Copernicus introduced the heliocentric system. More recently, the European Space Agency has referred to the Earth-Moon system as a double planet.

Indeed there are other reasons to regard the Moon as a planet, and the Earth-Moon system as a double planet. The Moon is not much smaller than Mercury, and is larger than the poor old ex-planet, Pluto. Further, the Earth-Moon mass ratio of about 1:80 is much larger than for any other planet-moon system, except for the Pluto-Charon pair. And, unlike all other planet-moon systems with the same exception of Pluto-Charon, the Earth-Moon system was created from a single planet by some sort of cataclysmic event, suggesting the system is one whole thing. So, the ESA's declaration of the Moon as half of a double planet may have been slightly whimsical but was definitely not groundless.

In any case, planet or other, your next Maths Masters column may be submitted from the Moon. And yes, we're aware of the risks: who knows the hostile aliens we might encounter? Still, we're pretty sure we won't encounter anything as hostile, or as alien, as can be found in Canberra.

Puzzle to ponder

If a coin rolls without slipping around another coin of the same size, how many times will it rotate while making one revolution?

[4]Imagine the Earth and the Moon playing on a giant seesaw. Then the center of mass is where you'd locate the pivot of the seesaw, so the two bodies would balance. See the discussion of Archimedes' lever, in Chapter 35.

CHAPTER 48

Tickling Orion with a triangle

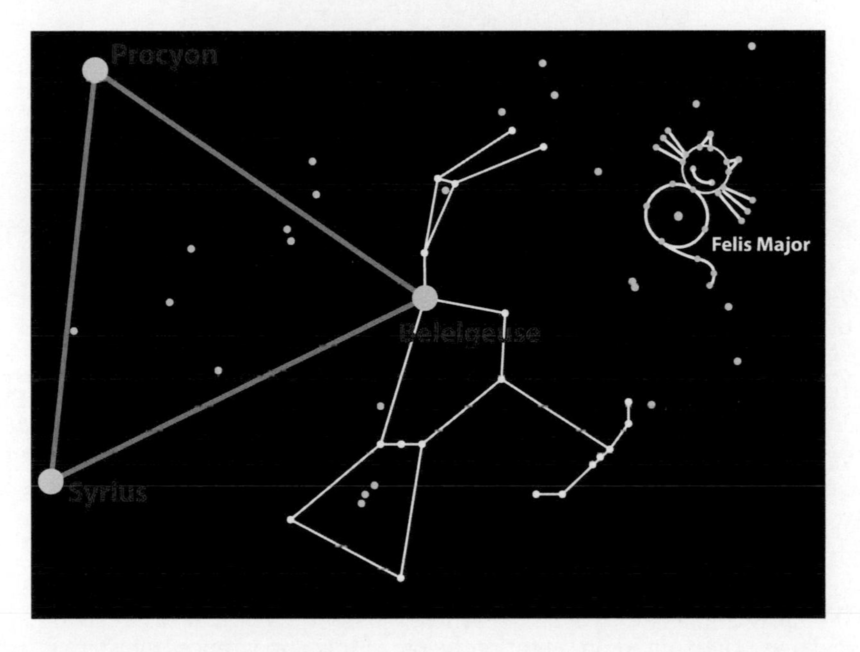

Regular readers of this column will be aware that your normally grumpy Maths Masters have been even grumpier of late. What, with the state of Australian Mathematics Education and the quality of our Glorious Political Leadership, we've even made plans to leave Earth and head to the Moon.[1] (Our many critics will be of the opinion that the sooner the better.) So, we've spent a lot of time recently, looking to the heavens.

We all know what happens when people scan the night sky for extended periods of time. They start seeing things that just aren't there: constellations in the shape of gods, cats, cows and so on. And what do your Maths Masters see? Definitely lots of cats, but in fact we see everything.

In a memorable scene from the 2001 movie *A Beautiful Mind*, mathematician John Nash (Russell Crowe) asks his girlfriend to name any object whatsoever. She opts for an umbrella and Nash immediately picks out a group of stars that form an umbrella-shaped constellation. She is very impressed.

[1]See the previous Chapter.

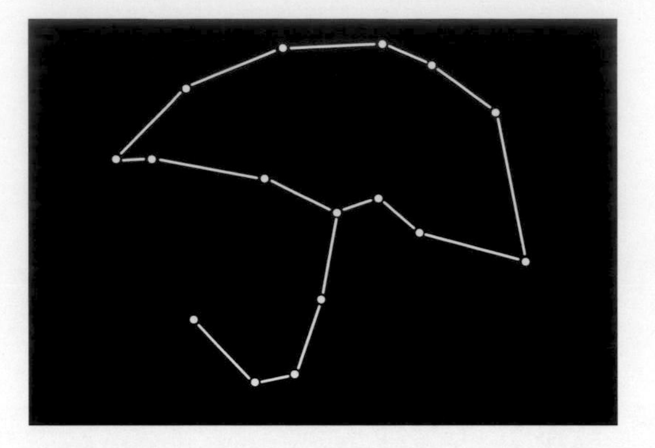

Your Maths Masters were much less impressed. Given the density of stars in the night sky it is pretty easy to find a star-to-star outline of almost anything. Moreover, general shapes such as umbrellas only have to be vaguely there to be recognizable.

Of course it gets a lot harder if we restrict ourselves to using bright and familiar stars, and if we attempt to locate a mathematically precise shape. There is in fact a constellation known as the Winter Hexagon, but we're still not overly impressed. It's not a *regular* hexagon and finding six bright stars that form a wonky hexagon is hardly a brilliant achievement.

Much more interesting is the Winter Triangle, consisting of Sirius, Procyon and Betelgeuse, three of the brightest stars in the sky. Our top picture shows the Winter Triangle, which appears very close to equilateral, tickling Orion's armpit. (To avoid having Orion stand on his head, we've shown the constellations as they appear in the Northern Hemisphere.)

An equilateral triangle of such humongous proportions is very impressive, even beating the triangular island we encountered previously.[2] Unfortunately, the Winter Triangle is not even close to being equilateral. It is actually very thin, much more extreme than our representation below. First appearances can be incredibly deceiving.

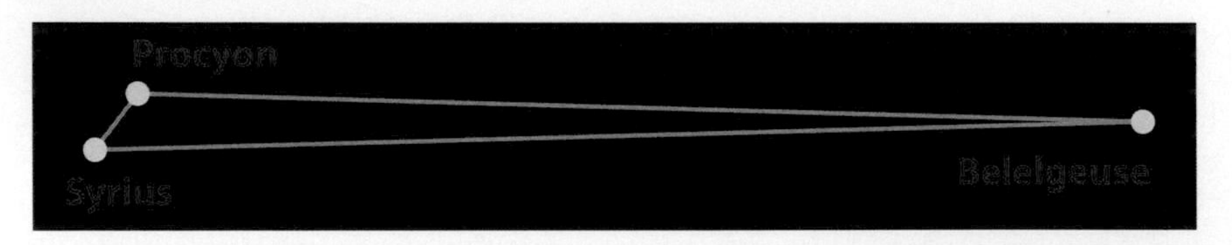

It's just luck that this particular star triangle appears to be equilateral when viewed from Earth. There is a fascinating mathematical truth, however, underlying such stargazing: for any star triangle there will be aliens on some suitably placed planet who will perceive that triangle as being equilateral.

There is a very nice mathematics activity - no spaceship required - that gives a good feel for this truth, and all this triangle viewing. Use some straws to make triangles of many different shapes. Then take one triangle at a time, move it around in space and view it with one eye closed. Try to orient the triangle until it appears to be equilateral. A closely related (though not equivalent) experiment can

[2]See Chapter 10.

be performed at Mathematikum, Germany's spectacular mathematics museum.[3] Below is a little maths mistress casting an equilateral shadow of, believe it or not, a very wonky triangle. Perhaps Scienceworks ... oh, never mind.[4]

After some such experimentation it is not hard to believe that any wonky triangle, when viewed from the correct angle, can appear to be equilateral. Proving that this can be done, however, is a little tricky.

Suppose we have a wonky triangle, ready for viewing. Set it aside, take a massive equilateral triangle and view this triangle straight on from a huge distance, as pictured below. The sight lines to the corners of the equilateral triangle will then form a thin equilateral prism. Our task now is to fit the wonky triangle inside the pyramid, with one corner lying on each of the sight lines.

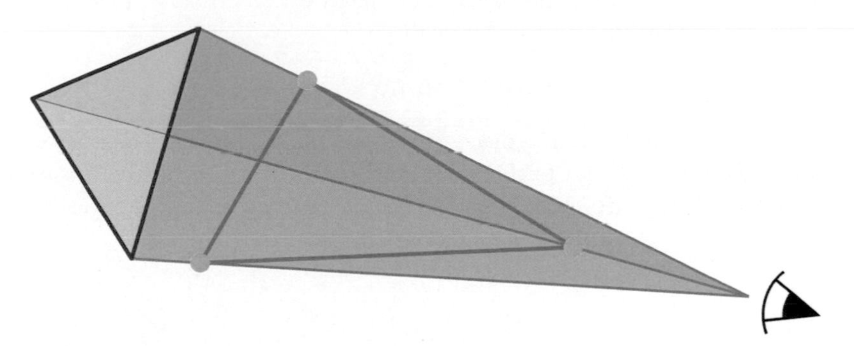

To see how this might be done, begin with the wonky triangle lying on one face of the pyramid, an edge along the bottom sight line and with a corner at the pyramid apex, as pictured below.

[3]See Chapter 5.

[4]Melbourne's not so spectacular science museum. See Chapter 40, and Chapters 63 and 64 of *A Dingo Ate My Math Book*.

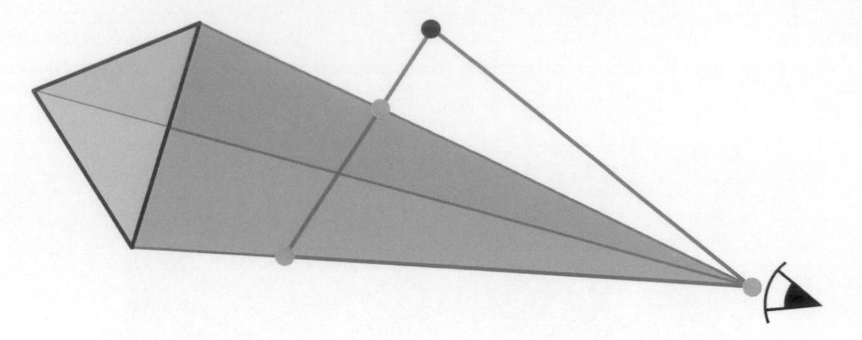

If we began with a sufficiently thin pyramid then the third corner of the wonky triangle will stick out beyond the pyramid. But that is easily fixed. We slide the other two triangle corners along the sight lines. If we slide far enough – a spaceship may now be required – the third corner will slide into the third sight line. Very pretty.[5]

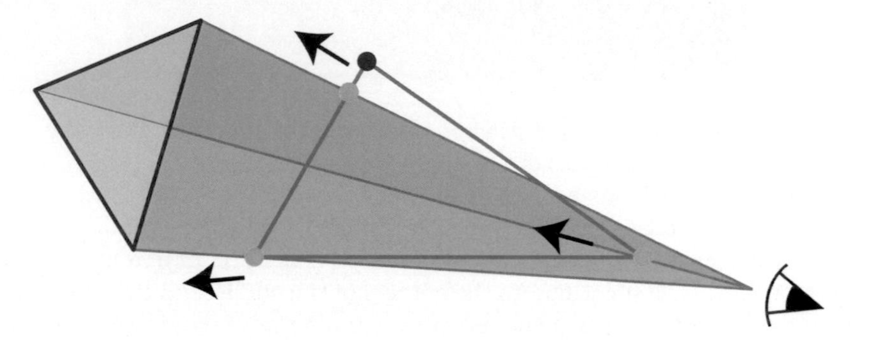

Some people look up to the skies and imagine umbrellas. Mathematicians, however, imagine beautiful proofs. And cats.

Puzzles to ponder

Cut an arbitrary triangle from a piece of paper. Show that with the rays of sun hitting a sufficiently large flat surface at right angles you can always position the piece of paper such that its shadow on the surface is an equilateral triangle.

[5]Proving that this can all be made to work is a little more difficult than it might appear. It can be done using the intermediate value theorem. See Chapter 22.

CHAPTER 49

The eternal triangles

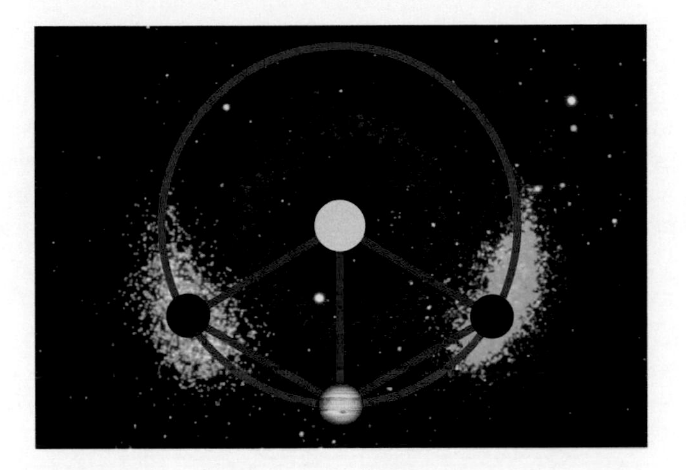

As part of our plans to flee to outer space, your Maths Masters have been pondering the geometry of the constellations. It appeared as if we had found an impressive equilateral triangle, but that turned out to be an illusion.[1] We have now located two genuine equilateral triangles, however, which may point us towards a new home.

The picture above shows many of the thousands of asteroids in the inner part of the Solar System. The red ring is the well-known asteroid belt lying between Mars and Jupiter. Further out are the so-called Trojan asteroids (green) and Greek asteroids (blue).

The Trojans and the Greeks share Jupiter's orbit and move in unison with this giant planet. The asteroids congregate around two central points, and together with Jupiter and the Sun these points form rotating equilateral triangles.

Very impressive, but what would cause two motley collections of asteroids to exhibit such perfect geometry? To figure out what's going on let's set Jupiter and the Sun rotating in (approximately) circular orbits around their common *center of mass*.[2] Now toss in a little asteroid and let's consider how it might react to the gravitational forces of the two big bodies.

---

[1] See the previous Chapter.

[2] This is the "lever point" between the two bodies. See Chapter 35

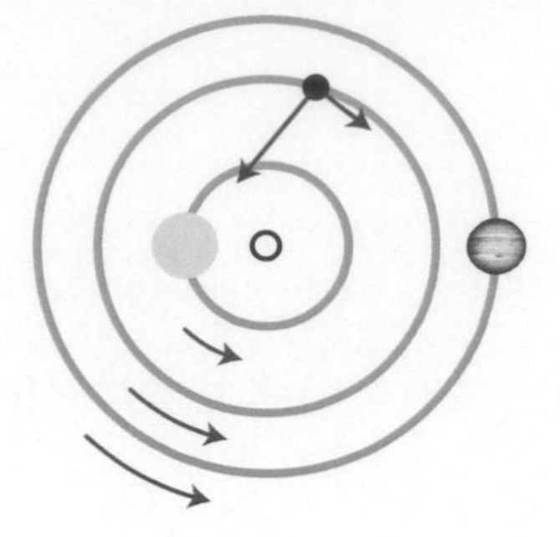

The asteroid may behave in a number of ways: it may fly off into the darkness of space; it may travel in a complicated orbit within the Sun-Jupiter system; if it's close enough to one of the two bodies then the asteroid may orbit that body as a little moon or crash into it. But finally, and much more interestingly, if the asteroid begins at a very special location and with just the right velocity then there the asteroid will be pulled by the exact *centripetal* force to keep it in limbo, also rotating around the center of mass and in a fixed position relative to Jupiter and the Sun.

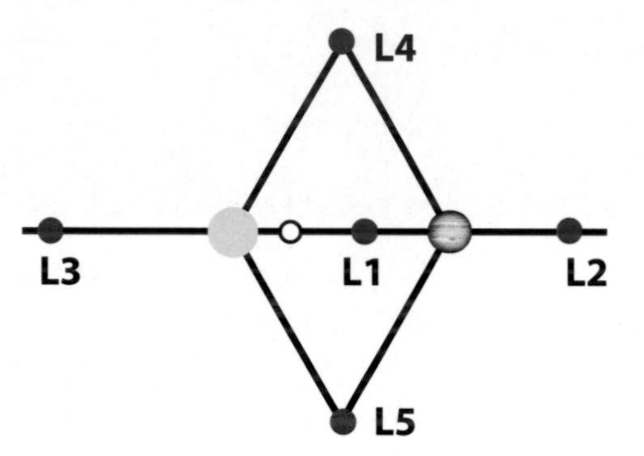

It turns out that this special alignment of forces occurs at precisely five locations, as pictured above. These locations are called the *Lagrangian points*, named after the famous French-Italian mathematician, Joseph-Louis Lagrange, who studied them in 1772.

Three of the Lagrangian points occur on the line connecting Jupiter and the Sun, and it is not too difficult to see why these equilibrium points should occur. Suppose, for example, the asteroid is between Jupiter and the Sun. If the asteroid is too close to either of two bodies that body's gravitational force will be too strong, and the asteroid will zip around more quickly. Somewhere between the extremes at either end, however, at the point labeled L1, there will be a happy medium; the combined forces acting at L1 will be just right to keep the asteroid at a fixed distance, orbiting around the center of mass with the just right *angular velocity*. It's all pretty intuitive, and the precise mathematics to back up the intuition is also reasonably straight-forward.[3]

Much more mysterious are the two remaining Lagrangian points, L4 and L5. There doesn't appear to be any easy intuition for why these equilibrium points

[3]The precise calculations are not difficult, but even without such calculations, the intermediate value theorem suggests why the point L1 should exist. See Chapter 22.

should exist, and locating them requires some serious calculation. Then, miraculously, these calculations show that L4 and L5 combine with Jupiter and the Sun to form the pictured equilateral triangles.

Why should those equilateral triangles appear? Anyone with a bit of university mathematics can work through the calculations, but it's all very unsatisfying. We're used to the idea that when a mathematical answer is so simple and symmetric then there must be a correspondingly simple explanation. For the Lagrangian points, alas, there appears to be no such easy intuition.

It is around L4 and L5 that the Greeks and Trojans congregate, and how they come to choose these Lagrangian points is also rather subtle. If an asteroid is merely close to but not exactly at a Lagrangian point then the gravitational forces will cause the asteroid to move relative to the two bodies. It turns out that if the asteroid is near L1, L2 or L3 then one of the gravitational forces will begin to dominate and the asteroid will be pulled away from the Lagrangian point. In the lingo, the Lagrangian points L1, L2 and L3 are *unstable equilibria*, similar to a ball at a saddle point between two hills. (The asteroid's behavior, however, is significantly more complicated.) In the Jupiter-Sun setting, however, with one body significantly larger than the other, L4 and L5 are *stable* equilibria. So, any asteroid that begins not too far from L4 or L5, and with a suitable velocity, will remain in the vicinity.

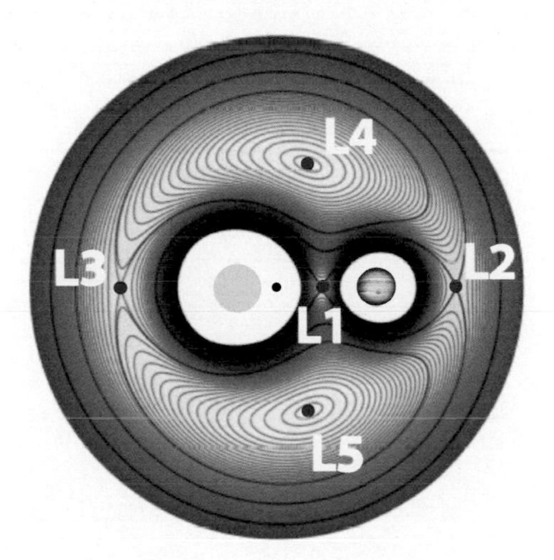

In general, a substantial space object trapped near the L4 or L5 point of two large bodies is called a *trojan*. Thousands of Jupiter-Sun trojans have been catalogued, but only a dozen or so other planet-Sun trojans are known, including just one Earth-Sun trojan. The L4 and L5 points of Saturn and its large Moon Tethys are occupied by the two smaller moons, Telesto and Calypso. Similarly, the L4 and L5 points of Saturn and its moon Dione are occupied by two smaller moons.[4]

What about the Earth-Moon system? Is there anything interesting hiding at their L4 and L5 points, perhaps a secret little moon? Disappointingly, there appears to be nothing but some dust clouds. Not so long ago, however, there were very grand plans to put something there.

[4]Mathematical physicist John Baez has a very good discussion of Lagrangian points and trojans, which, at the time of writing, is still available online.

The *L5 Society* was founded in 1975, with the mission to establish huge space colonies at the Earth-Moon L5 point. The idea was to manufacture solar powered satellites from lunar or asteroidal resources, so providing immense amounts of environmentally safe energy for Earth. For a short while the L5 Society was popular and influential. But, given the massive difficulty and cost to establish such a colony, interest soon waned.

Perhaps it's time to reconsider an L5 colony, and your Maths Masters are clearly ready to put their hands up to be part of such a venture. Or, even better, we can suggest a few Australian politicians to make the trip. They may be much less bothersome if they were inhabiting L5 instead of Earth.

Puzzle to ponder

The Sun has a mass of 2×10^{30} kilograms, Jupiter has a mass of 2×10^{27} kilograms, and the two bodies are 8×10^{8} kilometers apart. How far is the center of mass of the Jupiter-Sun system from the center of the Sun? Why is this interesting?

CHAPTER 50

On primes and Pluto

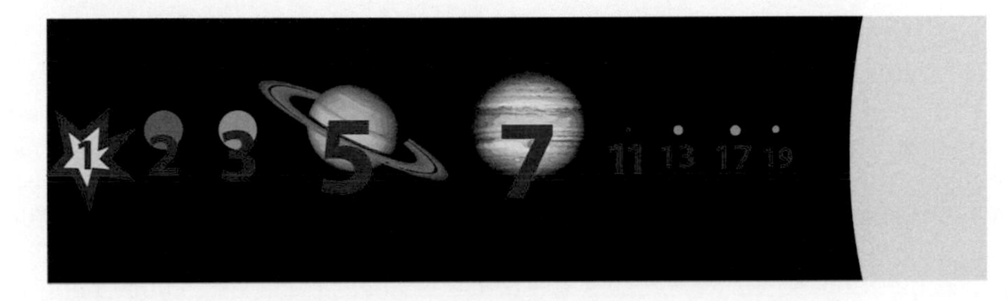

Is 1 a prime number? Not according to the Australian Curriculum, which bluntly declares that, whatever else, a number must be greater than 1 to be prime. So, that settles it. We'll be back next week.

Except, of course, such an answer settles nothing. As many a curious math student has noted, the opposite question is the follow-up: *why* is 1 not prime? After all, it is clearly not a product of other numbers. However, the frequent reply is a dismissive "because your teacher (and the Australian Curriculum) said so". Great help.

There are variations on the don't-bother-me response. For example, your Maths Masters are often apprised that a prime is a natural number – that is, a positive whole number – with exactly two distinct factors, namely 1 and itself; that leaves us with the expected 2, 3, 5, 7, 11 and so on. However this is nothing but legal shenanigans: the contrived qualification "exactly two distinct factors" has been inserted precisely to exclude 1, with still no justification for doing so.

And the justification can't be all that simple. After all, the number 1 *used* to be prime. Really.

	1
1	1
2	2
3	3
4	5
5	7
6	11
7	13
8	17
9	19
10	23

Just over a century ago, the mathematician D. N. Lehmer compiled his List of Prime Numbers from 1 to 10,006,721. And there, proudly occupying first place, is the number 1.

But maybe Lehmer was a one-off, 1-loving crank? Decidedly no. In the centuries prior to Lehmer many tables of prime numbers were published and some, although not all, began with the number 1. Clearly some explanation is in order, and the explanation begins on Pluto.

Here's the question: is Pluto a planet?

The answer, as most readers will know, is that Pluto is not a planet but that it used to be. Pluto was discovered in 1930 and was proclaimed to be the ninth planet in our solar system. Astronomers, however, always recognized that Pluto is very different from the other planets, and in 2006 the International Astronomical Union reclassified Pluto as a lesser, "dwarf planet".

The point is that the notion of "planet" is not God-given and it is not set in stone. Rather, astronomers *define* what they mean by the word. Then, as astronomers learn more about just what kind of things are zooming around out there, they refine their classifications, making "planet" more precise and more useful. And so, as it happens, Pluto is demoted.

The question of which numbers are prime is analogous. We cannot hope to *prove* that 1 is or is not prime; it is simply a question of how mathematicians have chosen to define "prime". And, though it is now accepted that 2 should be the first prime number, historically mathematicians have been neither clear nor consistent.

So how did mathematicians come to agree to exclude 1 as a prime, and why did it take them so long to do so? The answers take us into some weird and fascinating history.

The importance of primes is that they are the building blocks of the natural numbers; any composite number (a number greater than 1 that is not itself prime) can be written as a product of primes. For example, 84 is the product $2 \times 2 \times 3 \times 7$. Moreover, except for changing the order of the factors, 84 can be written as a product of primes in just that one way. That the same is true for any composite number is the very important (and not so easy to prove) *fundamental theorem of arithmetic.*

What if we permitted 1 to be prime? In that case, 84 would also have, for instance, the "prime" factorization $1 \times 1 \times 1 \times 2 \times 2 \times 3 \times 7$. That is, 84 could still be factorized, but it would no longer have a unique prime factorization.

The upshot is, if 1 is a prime number then describing prime factorizations is more complicated. That would seem sufficient reason to exclude 1 as a prime, and it is the reason your Maths Masters have always accepted. Both mathematically and historically, however, that reason turns out to be somewhat wide of the mark.

Consider again the fundamental theorem of arithmetic. Is it such a big deal if we're forced to replace "product of primes" with "product of primes greater than 1"? Hardly, and no one ever really considered it so. Indeed, until relatively recently the question barely even arose.

The fascinating history of such questions is documented in a wonderful paper by mathematicians Chris Caldwell and Yeng Xiong.[1] Quoting another excellent

[1] *What is the Smallest Prime?*, Journal of Integer Sequences, **15**, 2012.

survey,[2] Caldwell and Xiong point out that prime factorization was not of any great interest before the 19th century. Yes, division by prime numbers was of practical importance, but not the complete factorization.

It all really dates from 1801, when the great Carl Friedrich Gauss gave the first explicit statement of the fundamental theorem of arithmetic. Then, mathematicians began thinking seriously about the structure of numbers, and of whole worlds of numbers. And, once they started considering more exotic number worlds, things got very confusing.

In particular, mathematicians discovered very strange worlds in which a "number" can be factorized into "primes" in fundamentally different ways. In order to figure out what was going on, mathematicians were forced to be extremely thoughtful and precise with their definitions. A critical component of these definitions was to get the distractions out of the way, to classify do-nothing numbers such as 1 into their own separate group. That was the real impetus for 1 to cease being prime, though it still took another century for the modern definitions to take hold.

That would seem to be pretty much the story, except there is one more, astonishing twist. From the 19th century on there was good reason to exclude 1 from the primes, but what about prior to that? Yes, a number of mathematicians classified 1 as a prime number, but before 1600 it was very uncommon to do so. Why? Because 1 wasn't a number.

We bet you didn't see that coming. Certainly your Maths Masters didn't. In their paper, and in an extensive companion survey coauthored by Angela Reddick and Wilfred Keller,[3] Caldwell and Xiong consider very carefully how mathematicians throughout history have thought of "number". It turns out that, at least as far back as Euclid, most mathematicians prior to 1600 considered 1 to just be there. By contrast, "numbers" were different, effectively created from 1 by addition. It was only with the emergence of decimals in the late 16th century that excluding 1 as a number began to seem arbitrary and unnecessary.

The amazing fact is, for most of its history 1 has not been a prime number, simply because it hasn't been a number. If it seems difficult figuring out what mathematics is about now, just imagine what it was like in times past.

Puzzles to ponder

Consider the world of numbers of the form $a+b\sqrt{2}$, where a and b are integers. So, for example, $3+\sqrt{2}$ and 7 are numbers in this world. Now, explain why $3+2\sqrt{2}$ and $3-2\sqrt{2}$ are like "ones" in this world. Also, show that 7 is *not* "prime" in this world. That is, show that 7 can be written as a product of two numbers of the form $a+b\sqrt{2}$, neither of which is a "one".

[2] A. Göksel Ağargün and E. Mehmet Özkan, *A Historical Survey of the Fundamental Theorem of Arithmetic*, Historia Mathematica **28**, 207–214, 2001.

[3] *The History of the Primality of One: a Selection of Sources*, Journal of Integer Sequences, **15**, 2012.

CHAPTER 51

The math of planet Mars

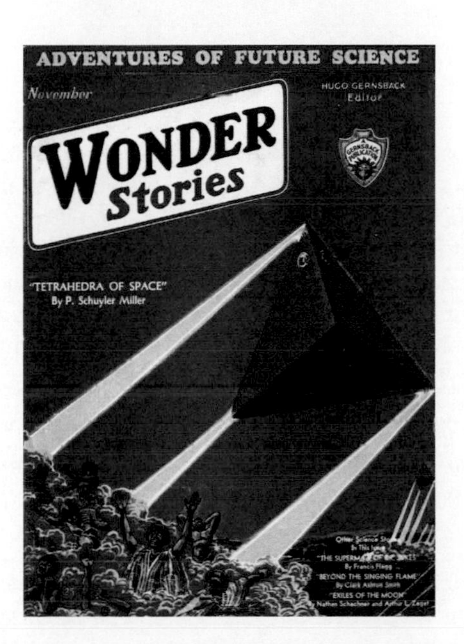

This year we're celebrating the Mathematics of Planet Earth, a great and noble effort to understand and to promote the essential role of the mathematical sciences in facing the challenges to our planet.[1] But what about other planets? If the Martians were similarly studying their home planet what might their mathematics be like?

Of course, imagining how Martians might think about mathematics, or anything, is definitely entering the realm of science fiction. It is not surprising, however, at least to fans of the genre, that science fiction can offer some genuine insight into mathematical fact.

One of the great early science fiction stories is *A Martian Odyssey* by Stanley G. Weinbaum. Written in 1934, it's a classic of the era of pulp science fiction.[2] The story is set in the 21st century, with astronaut Dick Jarvis trekking back to his spaceship and along the way meeting all manner of exotic Martians. There's no attempt at deeper meaning: it is simply intended to be, and is, great fun.

[1]MPE is a large, worldwide campaign, begun in 2013, an attempt to convince the general public that climate change is somewhat important, and that maybe we should do something about it. The campaign has worked a treat.

[2]The story appears in the stellar collection, *The Science Fiction Hall of Fame*, Doubleday, 1971.

Weinbaum, however, put plenty of thought into creating his weird creatures, and his story is concerned less with Martian actions than Martian thought. The result is a wonderful encounter of alien minds, an encounter in which simple mathematics plays a central role.

The main Martian character, and the real hero of Weinbaum's story, is Tweel, an ostrich-like creature that Jarvis saves from a tentacled monster. Consequently Tweel and Jarvis become friends and then attempt to communicate. Tweel learns a few English words but Jarvis can make no sense of Tweel's sounds, is simply unable to figure out how Tweel thinks. Then, Jarvis gets an idea:

> *"After a while I gave up the language business, and tried mathematics. I scratched two plus two equals four on the ground, and demonstrated it with pebbles. Again Tweel caught the idea, and informed me that three plus three equals six."*

That gave them some form of communication, if only just: *"We could exchange ideas up to a certain point, and then – blooey!"* Tweel, however, turns out to be very inventive with the little mathematics they share.

Tweel accompanies Jarvis on his trek and they come across a rock creature: the creature repeatedly builds a pyramid about itself, breaks out and continues on to build the next, larger pyramid. Weinbaum makes no attempt to explain this fantastic creature. Tweel simply summarizes with *"No one-one-two. No two-two-four"*.

Soon after, Jarvis begins to hallucinate and awakens to find himself entwined by a second tentacled monster. This time it is Tweel's turn to save Jarvis, after which he explains the monster as best he can: *"You one-one-two, he one-one-two."*

Finally, Jarvis and Tweel run into an army of barrel-shaped creatures, busily building mounds for no apparent reason. Tweel's description: *"One-one-two yes! Two-two-four no!"* That pretty much ends the story; there's a final confrontation with the barrel creatures, after which Jarvis is rescued by his fellow astronauts.

Weinbaum's story is wonderfully entertaining precisely because he makes no attempt to explain his fantastic creatures. We have the vivid impression that Tweel is thinking but without any real sense of how. And, Tweel is conveying *something* about the way the barrel creatures think but the reader is still as perplexed as Jarvis.

Obviously science fiction has no shortage of alien creatures, and if they're not simply monsters to be fought then there is usually the issue of communication. Of course it is easiest to just have the aliens speak English, but occasionally the issue is addressed. The writer imagines, as did Weinbaum, that aliens may not just think but also think dramatically differently to us earthlings.

So the question is, how does one communicate something, anything, in the most reliable manner? Weinberg's answer is compelling: mathematics offers the clearest, most fundamental and universal form of truth. If our alien cannot understand that $1+1=2$ then it is difficult to imagine that there can be any meeting of the minds.

But then how far does it extend? Would an alien that understands $1+1=2$ automatically know that $2+2=4$? What about prime numbers or Pythagoras or π? It's an interesting and instructive exercise to consider just how deep and how natural are our mathematical ideas.

In the 1997 movie *Contact*, which is based upon Carl Sagan's novel, the aliens' use of prime numbers gives the clue that their radio signals are not naturally occurring. That is very clever, even if the scriptwriter flubs the critical scene by having Jodie Foster refer to the primes as "base ten numbers". More amusing is the attempt to communicate with the aliens in the 1952 movie *Red Planet Mars*, by sending out the digits of π; it is perhaps reasonable to presume the Martians know of π, but sending π's digits is perhaps not the most clearly thought out plan. Below we've pictured a two-limbed Martian, pondering his own preferred expression for π.[3]

11.0010010001...

Of course considering how to communicate with aliens is not just the province of science fiction; there is no shortage of SETI fans pondering the problem. The *Arecibo message*, Earth's most famous hello to the Universe, is intimately mathematical. The message consists of a string of 1679 0s and 1s, the number 1679 chosen because it is the product, 73×23, of two primes. The idea is that our alien friend out there will be prompted to arrange the numbers in a rectangle. The resulting picture then indicates fundamental facts about numbers, chemistry, us earthlings and our solar system.

[3]Another π-loving alien appears in Chapter 2.

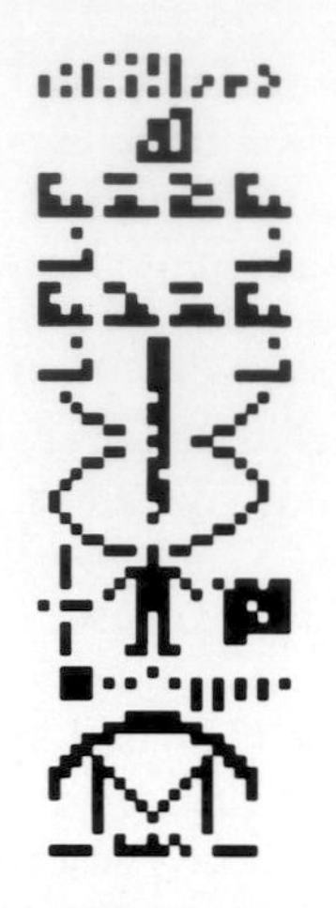

Undoubtedly the most elaborate work contemplating communication with aliens is LINCOS. This was an artificial language created in 1960 by Dutch mathematician, Hans Freudenthal.[4] Freudenthal literally begins with the whole numbers and $1+1 = 2$, works his way through elementary mathematics, on to language and human behavior, and ends with a description of planetary motion and our solar system. It is a remarkable and hilarious achievement; Freudenthal's 200 page message may be suitable for a curious and very patient alien but is absolutely unreadable by a human being.[5]

So what to make of it all? How should we really try to talk to our potential alien friends? Lord knows. But, mathematics seems as good an idea as any. And perhaps we can beam Stanley Weinbaum's wonderful story to the heavens: any intelligent alien out there should definitely appreciate that.

Puzzles to ponder

Why is Jodie Foster's "base ten numbers" line a goof?

Before making too much fun of dumb aliens, perhaps we should ponder: *why* does one plus one equal two? And, *why* does two plus two equal four?

---

[4] *Lincos, Design of a Language for Cosmic Intercourse*, North-Holland, 1960. Hans Freudenthal is more famous for establishing the Freudenthal Institute, and the birth of constructivism. We dearly wish he had stuck with talking to aliens.

[5] LINCOS makes an appearance as "Mathmatico" in an episode of the 70s TV series *Kolchak: The Night Stalker*. Unfortunately, those particular aliens are not interested in talking, and it doesn't end well. Those interested in such mathematics spotting may be interested in our book *Math Goes to the Movies*, John Hopkins University, 2012.

CHAPTER 52

Letter from Germany: The eternal grind

Maths Master Burkard recently visited *Phæno*, the amazing new science museum in Wolfsburg, Germany. It features many spectacular exhibits, including kilometer-long marble roller coasters and a double-storey fire tornado. One standout exhibit, however, seems to do almost nothing.

Machine and Concrete is a kinetic sculpture by the American artist Arthur Ganson. It consists of a series of twenty-four gearwheels, each with 120 teeth on the outside and fourteen teeth on the inner cog. A motor (on the right in the picture above) drives the first gearwheel at 9.24 rotations per minute, which then spins the second wheel, and so on down the chain.

The first wheel takes just under 6.5 seconds to complete a revolution. Then, because of the gear ratio, the second wheel takes $6.5 \times 120/14$ seconds – about 56 seconds – to complete a revolution. Multiplying again by $120/14$, we find that the third wheel rotates once every 477 seconds, and so on.

So what about the final gearwheel? Multiplying by $120/14$ a total of 23 times, it works out that the final wheel will take 18,800,000,000,000,000,000,000 seconds to make a revolution. That's one turn every 594 trillion years. Burkard decided not to wait around for that.

Such phenomenal numbers arise from time to time, but can we get a sense of how phenomenal they really are? Ganson's sculpture may help. Panning to the end of the sculpture, we can see that the last gearwheel is embedded in a concrete block; the final wheel is so close to stationary, the concrete makes no difference.

The concrete certainly makes for a very impressive punch line. Movies of *Machine and Concrete* can be found online, as well as videos of Ganson discussing his many amazing creations.

Let's now consider the gear arithmetic a little more carefully. Were twenty-four gearwheels really necessary? Below is the timeline that accompanies Ganson's sculpture, associating the time it takes for successive gearwheels to complete one revolution with events in the past:

1	6.5 seconds	
2	56 seconds	
3	8 minutes	
4	1 hour	
5	10 hours	
6	3.5 days	
7	1 month	
8	8.5 months	
9	6 years	
10	50 years	Half a human life
11	440 years ago	Copernicus's solar system
12	3,800 years ago	Stonehenge was built
13	30,000 years ago	Neanderthals died out
14	275,000 years ago	Woolly mammoths appear
15	2.4 million years ago	The beginning of the Stone Age
16	20 million years ago	The Himalayan Mountains appear
17	175 million years ago	Archaeopteryx appears
18	1.5 billion years ago	Multi-cellular organisms appears
19	13 billion years ago	The universe appears
20	100 billion years ago	
21	1 trillion years ago	
22	8 trillion years ago	
23	70 trillion years ago	
24	594 trillion years ago	

Clearly there is some overkill, and Ganson could have definitely made do without the last few gears. Depending upon the flexibility of Ganson's apparatus, we'd guess that thirteen gears would suffice for many years, and fourteen gears should last a lifetime.

Ganson's is definitely not the first attempt to give a sense of large numbers and the explosiveness of exponential growth. Perhaps most famous is *Powers of Ten*, the brilliant 1977 film by Charles and Ray Eames.[1] And, for kids, even very big ones, there is the wonderful book *How Much is a Million?*[2]

We'll close with the beautiful poem that accompanies Ganson's *Machine and Concrete*:

Æons of work
Divided, divided again
And still divided again ...
All stored effortlessly
Into microscopic spaces
Between teeth.

Puzzle to ponder

Paper is about 1/10 of a millimeter thick. Suppose you fold a piece of paper in half, then fold it again, and imagine you can keep folding indefinitely. How many times would you have to fold the paper for the folded stack to reach the Sun, 150 million kilometers away?

[1] *Cosmic Voyage* is a 1996 remake.
[2] David Schwartz, Collins, 1993.

CHAPTER 53

Calendar kinks

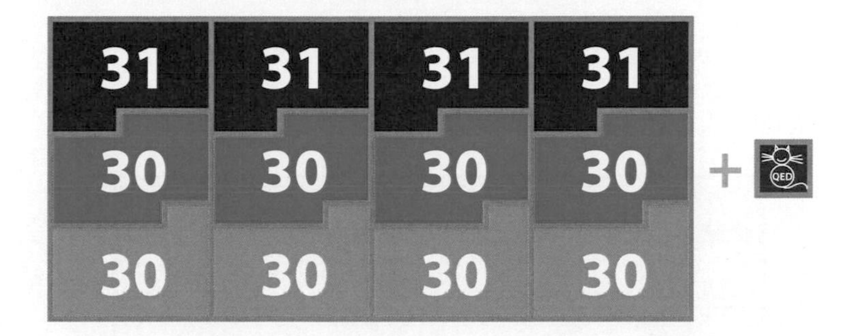

One of the fiddly little jobs of a university lecturer is to timetable the subjects he is to teach, the assignments and exams and so forth. The plan of course is simply to copy the previous year's timetable, and invariably the plan fails dismally. Alas, there is no consistency to when terms start and end, nor to how public holidays will fit in. Easter will do its crazy Easter thing,[1] and so on. It is not just annoying for lecturers and students; these date games befuddle everyone.

Wouldn't it be convenient to have a perpetual calendar, one that simply repeats itself like a clock? Then any special occasion, such as last Friday's Anzac Day,[2] would always be on the same day of the week. School terms and holidays could always start and end on the same dates, and on and on. No more guessing of dates and no more juggling to fit in with the vagaries of each new year's calendar.

Of course such a calendar must be an impossibility. We wouldn't have suffered through millennia of calendric randomness if there were an easy solution, right? Wrong.

There are a number of simple and clever perpetual calendars, any of which would be a vast improvement on our current clunky one. So why are they so secret and why can't the World agree to adopt one of them?

Any calendar begins with the observation that a year, the time it takes the Earth to complete its trip around the Sun, is about 365 days. So, we might simply number the days from 1 to 365 and be done with it.

Very simple, if not very useful. We would also hope to divide our year into something like weeks or months or seasons, as we currently do. That's where we can easily be led astray.

If we want to stick with tradition and declare a week to be seven days it seems we're doomed to a calendar that changes each year. 7 does not go evenly into 365

---

[1] See the next Chapter.

[2] Celebrated each April 25, Anzac Day is essentially Australia's version of America's Memorial Day, to remember those who served and died in wars.

and that's that. Months, traditionally based upon the Moon's orbit of the Earth, are even trickier. The Moon's trip takes about 29 1/2 days. We can crib a bit and go for either 29-day or 30-day months, but neither goes neatly into 365.

Perhaps we can rethink some of this. After all, there is nothing God-given about a 7-day week. Well, as it happens, the 7-day week *is* God-given, or at least religion-given. But that's not a compelling argument for a modern world. And there is hardly a compelling call for lunar months, except amongst werewolves.

If we're happy to ditch the traditions then we're left with figuring out how to divide 365 into smaller, equal units. Unfortunately, $365 = 5 \times 73$, and 5 and 73 are both prime numbers. That means our only options are five "months" of 73 days each, or seventy-three "weeks" of 5 days each. Neither is very appealing.

Let's cheat. if we remove a day then we're left with $364 = 13 \times 28$ days, which is much more promising. We can then have thirteen months of 28 days, and each month can be divided into four weeks of 7 days.

Sun	Mon	Tue	Wed	Thu	Fri	Sat
1	2	3	4	5	6	7
8	9	10	11	12	13	14
15	16	17	18	19	20	21
22	23	24	25	26	27	28

This calendar resembles our existing one but is much more regular. Not only would every year start on the same day of the week, so would every month. It could also have the spooky bonus of providing us with a Friday 13th every month. Easter, alas, would probably continue to do its crazy Easter thing, but there's not much we can do about that.

That'd be great but what about the missing, 365th day? That's easy: we just chuck it in at the end of the year as an extra holiday, with no day of the week assigned to it. We suggest that it be named QED Cat Day.

Of course there's that whole leap year thing, where we account for a year being a bit longer than 365 days, ensuring that our calendar stays in synch with the seasons. That can be handled just as it is now, adding an extra leap day to a year as required, more or less once every four years. Again, the trick is simply to not assign a day of the week to this extra day.

This is a promising calendar but admittedly having thirteen months in a year could be considered unsatisfactory. It would be convenient to have the calendar mark off seasons of equal length and 13 is another annoying prime number. Luckily, there is a simple alternative that will work.

We can factorize $364 = 4 \times 91$, which means we could have four seasons of 91 days each. Then each season could consist of three months of 31, 30 and 30 days. That would be very similar to our current calendar of twelve months, with QED Cat Day again included to complete the year.

These are great ideas but QED Cat doesn't deserve (and is unlikely to get) all the credit. Such calendars have been around a while and have already received significant attention. The idea of a 13-month calendar dates back to the 18th century and our version above, known as the International Fixed Calendar, was seriously considered for adoption by the League of Nations in 1923. The perpetual 12-month calendar above, known as the World Calendar, was considered by the United Nations in 1955.

Why weren't the calendars adopted? It seems that God will giveth clumsy calendars but God won't taketh them away. The failure of these calendars to win approval was mainly due to the opposition of religious fundamentalists. Though weeks would have still consisted of the traditional seven days, that one extra QED Cat Day was deemed to violate the "rest on the seventh day" thing. Sheesh.

These days calendar reform is less of a concern. We all have computers to help us cope with the consequences of pious obstinance. Still, a sensible calendar would be pleasing and is well overdue. And, with the steady waning of religion's intellectual (and moral) authority, who knows what the future will bring?

Actually, what the future will definitely bring is a year of 364 days: the years are slowly but inevitably getting shorter and the days are getting longer. In a few million years or so, the year will have lost a day and finally it'll be possible to please everyone. For a while. Until we lose another day, and we have to begin the haggling all over again.

Puzzle to ponder

Suppose you decide that lunar-ish months are really important, so you demand equal months of 30 days each. How might you construct a calendar to accommodate this?

CHAPTER 54

Strange moves of a mathematical feast

The eggs above contain the dates of Easter Sunday from 2006 to 2009. With such an obvious pattern, it is clear that Easter next year will be ... who knows!

Well, we do. But it was work. The precise rules for determining Easter are strange and intricate. And, trying to find a pattern to the Easter dates is anything but easy.

The date of Easter was meant to have been settled by the Catholic Council of Nicaea in 325. In one big mouthful, they declared Easter to be on the first Sunday after the first full moon, on or after the vernal equinox – the moment at which the Sun moves from the northern to the southern hemisphere. This is no longer true.

Why? That's not so obvious. After all, with the Nicaean concept of Easter it is easy to calculate year by year. For example, this year the equinox was March 20, followed by a full moon on April 9, and then Easter on April 12.

What about next year? Starting at April 9, we mark off time to next year's equinox, again on March 20. We just count in lunar months, from full moon to full moon, about 29.5 days each. This should get us to a full moon on March 29. Then the following Sunday is April 4, and that's Easter.

That seems clear and simple to us. Not, apparently, to the Christian scholar Dionysius Exiguus. In 525, Dionysius changed the definition of Easter.

What's the problem? Well, a minor problem is that lunar months vary in length. This means our calculation above was a bit imprecise. And in fact the full moon next year will occur on March 30, though still leading to Easter on April 4. This is really a non-issue: one can easily be more precise if the timing of the moon warrants it.

It seems that the real issue for Dionysius wasn't the calculation of Easter for the next year or two. His concern was to calculate *all* the Easters! Why did he want to do that? God only knows.

Dionysius standardized the dating of Easter, using the Julian calendar of the time. First, he declared the "equinox" to be March 21. So, he no longer needed to bother with actual equinoxes. Second, Dionysius replaced the calculation of an actual full moon with a theoretical calculation, the *Metonic Cycle*. Named after the Greek astronomer Meton, this is the observation that 235 lunar cycles takes extremely close to 19 years. Dionysius simply declared 19/235 of a year to be a lunar month and used this to count off. Choosing an actual full moon on which to begin, Dionysius then happily calculated the dates of hundreds of "moons", and hundreds of Easters.

Alas, Dionysius used a poor calendar. The Julian calendar included exactly one leap year every fourth year. This was inaccurate, and the Julian calendar fell seriously behind.

The problem was substantially corrected in 1582 by the Gregorian calendar, which is still used today. The key was to eliminate certain leap days, in those years divisible by 100 but not by 400. So, with the length of a year settled, can we now continue with Dionysius's method? Nope. There's more.

Pope Gregory did not simply adjust leap years. His scholars also knew that the Metonic Cycle was not exact, and so this was adjusted in a very similar manner. In effect the dates of the full moons were shifted 8 days every 2500 years. With that final adjustment, we have the currently sanctioned method for calculating Easter.

All of the above is precisely mathematical, and so it can all be written out as one big set of equations. The great mathematician Karl Friedrich Gauss was probably the first to explicitly do so. Compiled into one monster equation, here it is:

$$\text{year} = \mathrm{Y}$$

$$\text{day} = \Bigg(\bigg(19Y_{19} + \left\lfloor \frac{Y}{100} \right\rfloor - \left\lfloor \frac{\left\lfloor \frac{Y}{100} \right\rfloor}{4} \right\rfloor - \left\lfloor \frac{\left\lfloor \frac{Y}{100} \right\rfloor - \left\lfloor \frac{\left\lfloor \frac{Y}{100} \right\rfloor + 8}{25} \right\rfloor + 1}{3} \right\rfloor + 15\bigg)_{30} + \bigg(32 + 2\left\lfloor \frac{Y}{100} \right\rfloor_4 + 2\left\lfloor \frac{(Y_{100})}{4} \right\rfloor - \left\lfloor \frac{\left\lfloor \frac{Y}{100} \right\rfloor + 8}{25} \right\rfloor - (Y_{100})_4\bigg)_7 - 7\left\lfloor \frac{Y_{19} + 11\left\lfloor \frac{\left\lfloor \frac{Y}{100} \right\rfloor + 8}{25} \right\rfloor + 22\left(32 + 2\left\lfloor \frac{Y}{100} \right\rfloor_4 + 2\left\lfloor \frac{(Y_{100})}{4} \right\rfloor - \left\lfloor \frac{\left\lfloor \frac{Y}{100} \right\rfloor + 8}{25} \right\rfloor - (Y_{100})_4\right)_7}{451} \right\rfloor + 114\Bigg)_{31} + 1$$

$$\text{month} = \left\lfloor \left(\bigg(19Y_{19} + \left\lfloor \frac{Y}{100} \right\rfloor - \left\lfloor \frac{\left\lfloor \frac{Y}{100} \right\rfloor}{4} \right\rfloor - \left\lfloor \frac{\left\lfloor \frac{Y}{100} \right\rfloor - \left\lfloor \frac{\left\lfloor \frac{Y}{100} \right\rfloor + 8}{25} \right\rfloor + 1}{3} \right\rfloor + 15\bigg)_{30} + \bigg(32 + 2\left\lfloor \frac{Y}{100} \right\rfloor_4 + 2\left\lfloor \frac{(Y_{100})}{4} \right\rfloor - \left\lfloor \frac{\left\lfloor \frac{Y}{100} \right\rfloor + 8}{25} \right\rfloor - (Y_{100})_4\bigg)_7 - 7\left\lfloor \frac{Y_{19} + 11\left\lfloor \frac{\left\lfloor \frac{Y}{100} \right\rfloor + 8}{25} \right\rfloor + 22\left(32 + 2\left\lfloor \frac{Y}{100} \right\rfloor_4 + 2\left\lfloor \frac{(Y_{100})}{4} \right\rfloor - \left\lfloor \frac{\left\lfloor \frac{Y}{100} \right\rfloor + 8}{25} \right\rfloor - (Y_{100})_4\right)_7}{451} \right\rfloor + 114 \right) / 31 \right\rfloor$$

In fact Gauss was not the first to indulge in such calculations. The Jesuit mathematician Christopher Clavius was instrumental in having Pope Gregory accept the new calendar.[1] Ever the optimist, Clavius computed Easter dates up to 80 million years into the future. Should we have told him that by that time the Gregorian calendar will be more than 20,000 days out of sync? Or that by then a year will be only 360 days long? Nah, we didn't want to spoil his fun.

Puzzle to ponder

Try to make sense of the crazy formula for the date of Easter.

[1] See Chapter 60.

CHAPTER 55

Lucky Friday the 13th

It's a spooky time. Halloween has just passed, and this week we have Friday the 13th, the third such Friday this year.[1] How unlucky is that! Or, how lucky, if you're a black cat or a demonised Maths Master.

Judged by this criterion, 2009 has been unusually spooky. By comparison, there was only one Friday the 13th in 2008, and there will be only one next year, in 2010, and one again in 2011.

How many do we expect in a given year? There are seven days of the week and twelve months of the year, and 7 into 12 won't go. In the long run, however, it's a fair guess that it will all even out, with a 1/7 chance of the 13th of a given month being a Friday. Surprisingly, this is not the case. In fact, vampires rejoice, it turns out that the 13th is most likely to fall on a Friday.

To see how this works, we have to review our calendar system. Recall that Australia, along with most countries, uses the Gregorian calendar. We've written of this calendar previously, when discussing the dates of Easter.[2] This calendar was introduced in 1582 and is based on a 400 year cycle. The cycle refers to the choosing of leap years, but it turns out that a 400-year period contains 146,097 days, which

[1] This column was written in 2009.

[2] See the previous Chapter.

comes to exactly 20,871 weeks. This tells us that the days of the week have the same 400-year cycle. For example, today is Monday November 9, 2009, and so the cycle tells us that November 9 in the year 2409 will also fall on a Monday.

This means that to determine the frequency of Friday the 13th, we just have to check through the 4800 months in a 400-year period, and count the number of spooky Fridays. Below is the frequency table for the 13th falling on the days of the week for such a 400-year cycle.

Sun	Mon	Tue	Wed	Thu	Fri	Sat
687	685	685	687	684	688	684

As you can see, Friday is the winner with 688 hits. But, the vampires shouldn't get overly excited: the chances of the 13th being a Friday is $688/4800 = 14.33\%$, as compared with $684/4800 = 14.25\%$ for Thursday or Saturday, the least likely days.

Though the calendar takes 400 years to cycle, there are only 14 possible calendars for a given year. Why? There are seven possible days on which January 1 can fall, and it is either a leap year or not: then, $7 \times 2 = 14$.

Day of the week Jan 1	Non-leap year Fri 13th	Leap year Fri 13th
Monday	April, July	September, December
Tuesday	September, December	June
Wednesday	June	March, November
Thursday	February, March, November	February, August
Friday	August	May
Saturday	May	October
Sunday	January, October	January, April July

By inspecting all 14 calendars we obtain the complete picture above. We see that every year must have at least one Friday 13th. And, with three such Fridays, this year was indeed the spookiest possible.

Puzzle to ponder

Which is the spookiest month? That is, which month most frequently contains a Friday 13th?

Part 7

On the Shoulders of Lesser Giants

CHAPTER 56

Hermann the hermit

Regular readers of this column may suspect that your Maths Masters hate decimals, and base ten and the metric system. It's not true. We don't actually hate decimals and the like; we're just happier when they're not around.[1] There is, however, one base ten phenomenon we can never resist: we love a good centenary. It's our guilty pleasure.

Each year we search for mathematicians who were born or who died a century ago, or two hundred years ago, and so on, back as far as we can see. It's a fun way to choose to honor some great mathematicians, and some of the long forgotten. And, you never know what mathematics this game of lucky dip will bring forth.

This year we have the opportunity to celebrate the *thousandth* birthday of German mathematician, Hermann of Reichenau. No, we had never heard of him either. Reading up on Hermann, however, gave us some fascinating insight into medieval mathematics.

Hermann of Reichenau wasn't exactly a hermit, as our title unfairly labels him, but he was a monk and he didn't get out a lot. Hermann was severely disabled from childhood, with little ability to walk or speak. He was untactfully referred to as Hermann the Lame, and even less tactfully as Hermannus Contractus.

[1]See Chapters 2, 51 and 59. It's not that we particularly love other bases, although we do. And, it is not that we particularly think the metric system is oversold, although it is. Our main concern is that the emphasis on decimal representation, hand in hand with the inevitable calculator-hawking, undermines the teaching of arithmetic and, most specifically and importantly, the teaching of fraction arithmetic.

In the 11th century, European mathematics was in very poor shape. The great achievements of the Greek mathematicians had been long forgotten, and the Indian and Arabic advances were only beginning to find their way to Europe. As such, European mathematics of the time consisted of little more than rudimentary astronomy and arithmetic. The arithmetic was based almost solely on the clumsy and semi-mystical investigation of number patterns by the sixth century Roman mathematician Anicius Boethius. The mathematics historian Florian Cajori described Boethius as "a [giant] among Roman scholars, and a [midget] by the side of the Greek masters". This was Hermann's role model, and those were the times.

Hermann created no original mathematics, but he was an industrious writer and he did his small bit to waken European mathematics from its slumber. Hermann was one of the first to write on Arabic mathematics and astronomy, and as such was responsible for the introduction of the astrolabe – a precursor of the sextant – and other astronomical instruments. It seems that Hermann could not read Arabic, most likely learning his mathematics from the writings of Gerbert of Aurillac, the greatest European mathematician of the era – and a Pope.

There's not much more to say about Hermann's mathematics; it wasn't thrilling. *Except*, we have Hermann to thank for introducing us to the craziest board game of all time: *rithmomachia.* Once rivalling chess in popularity, rithmomachia is unbelievably complicated. Hermann of Reichenau was responsible for one of the earliest known manuals for the game, and it definitely required one.

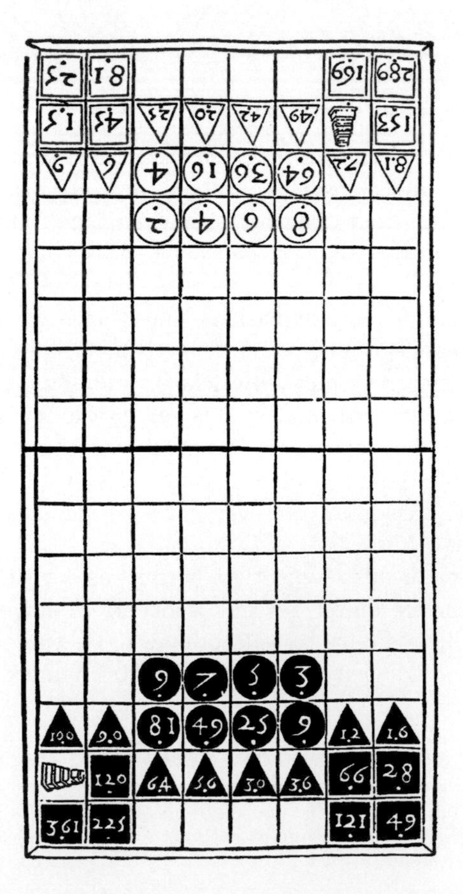

"Rithmomachia" translates literally as "the battle of number", which is as good a description as any of the game. The expression "mathematical Calvinball" also comes to mind. The rules varied, but typically the game was played on an 8×16 board, the players beginning with numbered counters, as pictured below.

Some patterns to the numbers are clear, even if their purpose is not, and the rest are easy to describe, if seemingly arbitrary. The 153 in the white square on the left, for example, comes from adding $81 + 72$ in the neighboring triangles.

The counters move horizontally and vertically, the same as rooks in chess, but they may only move a specific number of spaces; circles, for instance, may normally only move one space. Counters can be captured in a similar manner to chess, by *meeting*. In the following diagram, for example, Black's circle 7 can capture White's square 15. There are other ways that counters can be captured, however, and that's where the fun, and the arithmetic, begins.

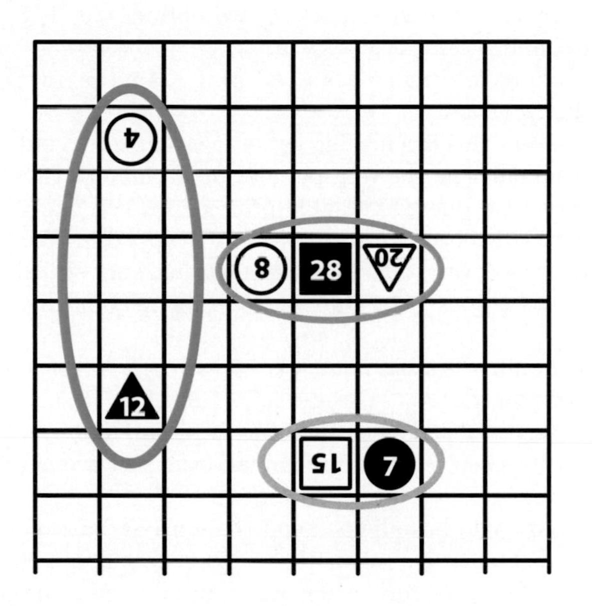

A second way to capture is *ambuscade*. For example, Black's square 28 above can be captured by ambuscade because it is sandwiched between White's 20 and White's 8, and $28 = 20 + 8$. What fun! But wait, there's more. In the diagram above, White's circle 4 can capture Black's triangle 12 by *assault*. Why? Because there are 3 vacant squares between the counters, and $4 \times 3 = 12$.

Now, take a deep breath; we have to describe how to actually win the game. There were many ways to win rithmomachia, and the players would agree prior to the game which type of win they were attempting. For example, the goal may be to capture a certain number of counters. Or, the goal may be to have the numbers on the captured counters sum to a designated value. Or, both may be required.

Those games, however, were for the novices. The true rithmomachia aficionado wanted to play for a *Proper Victory*. These were obtained by arranging counters in a number sequence. For example, arranging the arithmetic sequence 6, 9, 12, involving both color counters, would constitute a proper victory. Then, among proper victories, one was prized most of all: a type of triple win called the *Excellent Victory*. An example of such a victory is illustrated below.

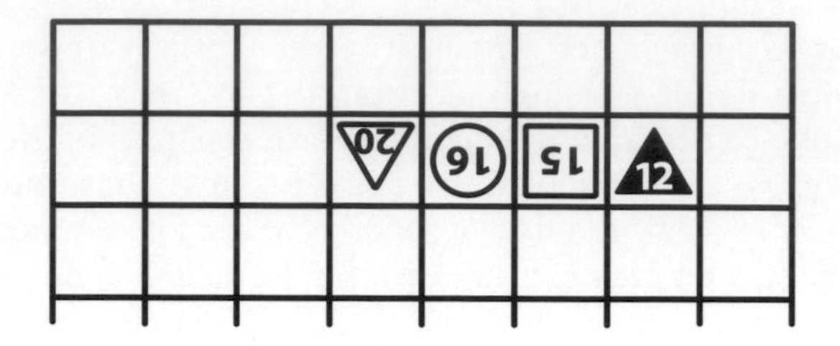

What, you can't see the excellence of this victory? Then let us explain. First, ignoring the 15, the numbers 12, 16, 20 form an arithmetic sequence. Secondly, $20/16 = 15/12$ are equal proportions, which was a second type of proper win. Finally, ignoring the 16 and dividing by 60, we obtain $1/5$, $1/4$, $1/3$, which is a *harmonic sequence*, and a third type of win.

There's plenty more to rithmomachia, all in the same vein.[2] So, anybody keen for a game? Anybody? Hmm.

It is no surprise that this astonishing game eventually died out, but it's also not surprising that rithmomachia was very popular in Hermann's time. Rithmomachia fitted perfectly with, and offered valuable practice in, the flawed art of medieval arithmetic.[3] In the 11th century, arithmetic was wrapped in Boethius's mysticism and, consequently, it was very difficult. Calculations were still done with Roman numerals and the techniques were clumsy. The memory of myriad special cases played a huge role.

Once the now-familiar Arabic numerals appeared, and once the power of positional notation was understood, arithmetic lost much of its mystery, and the era of rithmomachia was set to end. The gameboard pictured above, featuring Arabic numerals, is from a 16th century French manual, from just around the time interest in the game was waning.

Indeed, Hermann could have helped end the era. Gerbart of Aurillac included Arabic numerals in his writings, but neither he nor Hermann recognized their true significance. That had to wait for a later, and much greater, mathematician.[4]

So, Hermann of Reichenau missed an opportunity to revolutionize European mathematics. But, he did leave us with one crazy board game. That's plenty good reason for us to celebrate him on his thousandth birthday.

Puzzle to ponder

We can't think of a puzzle. Make up your own, and then answer it.

---

[2]See D. E. Smith and C. C. Eaton, *Rithmomachia, the Great Medieval Game*, American Mathematical Monthly, **18**, 73–80, 1911.

[3]See Ann Moyer *The Philosophers' Game*, The University of Michigan Press, 2001. Moyer's book delves deeply into the role of rithmomachia in Medieval and Renaissance education.

[4]The famous Fibonacci didn't actually come up with "his" numbers; see Chapter 58. Fibonacci was instrumental, however, in the popularization of Arabic numerals in European mathematics. See *The Man of Numbers* by Keith Devlin, Bloomsbury, 2011.

CHAPTER 57

The equals of Robert Recorde

You may not have heard of Robert Recorde. We previously knew little of the 16th Century mathematician, although his name arose briefly in our discussion of *surd*, Recorde providing the earliest known English usage of the word.[1] But, this year is the 500th anniversary of Recorde's birth, and this rare milestone gives us a good excuse to highlight Robert Recorde's notable achievements.

To begin, we should confess that our, and others', celebrations may be premature. Robert Recorde is known to have been born in the Welsh town of Tenby, and his birth year is nearly always given as 1510. But it will come as no surprise that there is slim evidence for this date. Indeed, there is reason to believe that Recorde was actually born in 1512 or 1513. No matter: we'll simply honor Robert Recorde again in a couple of years.

Robert Recorde was the first notable writer of mathematics in English. Recorde's *The Ground of Artes*, published in 1543, was England's first popular arithmetic manual. It was followed by Recorde's 1551 textbook on geometry (in which he first used the word "surde"), his 1556 textbook on astronomy, and finally his 1557 textbook on algebra, *The Whetstone of Witte*. Recorde then died in prison in 1588, the consequence of his initiating a legitimate but ill-advised legal action against a powerful enemy.

Though educated at Oxford and Cambridge, it was no accident that Recorde chose to write in English, rather than the standard scholarly Latin of the time.

[1] In our *Age* column *Theatre of the surd*, we looked at the history of "surd" and tried to determine whether the word ever meant anything more precise than "rooty thing". It seems pretty much not.

Recorde's focus was always upon clarity and practicality. He emphasised applications, though with respect for the deeper, underlying ideas. Recorde's texts are now difficult to read, but in his time they were considered models of thoughtful, careful exposition, and they remained in print long after his death. *The Ground of Artes* went through a large number of editions, up until 1699.

As an early English writer of mathematics, Recorde had many opportunities to introduce new expressions for mathematical quantities. Many, such as the remarkable *zenzizenzizenzike*, indicating the eighth power of a number, were clumsy and, unsurprisingly, have not survived. When based upon simple English, Recorde's invented terminology was more natural. He later ceased this practice, deciding that contrived English expressions would be "obscuryng the olde Arte with newe names". He was perhaps correct, though we're really disappointed that his *like flattes* has not survived, to describe similar (i.e. same shape) plane figures. But, at least Recorde's *straight line* is still going strong.

Undoubtedly, Robert Recorde is most famous for having introduced "=", our modern equals sign. It appears in "The Whetstone of Witte", Recorde's first ever equation being $14x + 15 = 71$:

It is no coincidence that Recorde's equals sign was very long. The symbol was already in common use in geometry, to indicate lines being parallel. Recorde simply adopted the symbol of parallel lines as a metaphor,

because noe .2. thynges, can be moare equalle.

Recorde's introduction of the equals sign was much more than the gift of an elegant symbol. Prior to Recorde, very few mathematicians had employed any such symbol, instead simply stating an equality in words. The introduction of the symbol permitted Recorde to emphasise the fundamental role of equality in algebra: if we take expressions for two equal numbers and if we then perform the same manipulations on the two expressions, the resulting numbers must still be equal. In Recorde's words,

if you abate even portions from thynges that bee equalle, the partes that remain shall be equall also.

Equality is the single most important concept in mathematics. Much too often, we witness writers shy away from using "equals", employing instead a cowardly and confusing "equivalent". By this and other means, there is a continual devaluing of the central role of equality. Recorde's equals symbol is powerful and beautiful, and deserving of much greater acknowledgment and respect. We proudly and humbly honor Robert Recorde, on the occasion of his 500th (or 498th) birthday.

Puzzle to ponder

Consider the equation: $(-1)^{\frac{2}{6}} = (-1)^{\frac{1}{3}}$. Is this equation true or false?

CHAPTER 58

Pythagoras's theorem ain't Pythagoras's

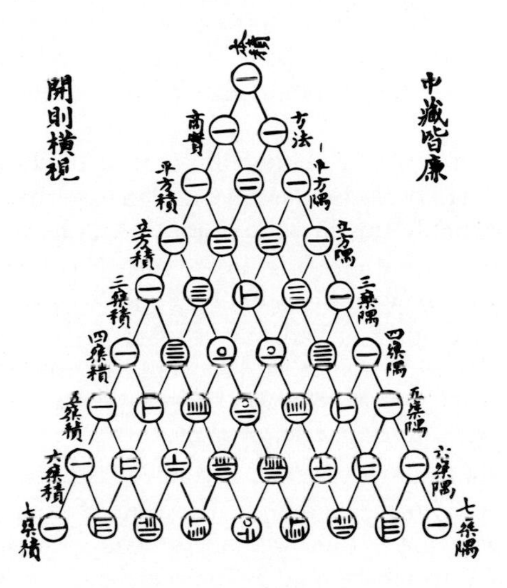

Since you're reading this column, you are probably a regular consumer of popular mathematics. If so, you may have read about Cambridge mathematician John Barrow and his recent contribution to competitive rowing. Barrow's breakthrough was to realize that the commonly used left-right arrangement of the oars causes a boat to wobble, and that certain other arrangements are theoretically more stable. His very elegant paper was widely reported.[1]

It is evident, however, that some reports have overstated the newness of Barrow's new idea. Barrow himself remarks that in the 1950s some Italian rowers had adopted a stable rowing arrangement, and that the rowing world has generally known of such arrangements for a long time. It seems that the rowers never isolated or articulated the underlying mathematical principle, but this was most definitely done in a 1977 paper by mathematician M. N. Brearley. And, in Australia, the ideas had been widely disseminated through the talks and writings of sports mathematician, Neville de Mestre.

We have no doubt that Barrow acted in good faith, and he references Brearley's and de Mestre's prior work in his paper. Moreover, Barrow's work is significantly more general than Brearley's and contains interesting new ideas. Nonetheless, through some inaccurate reporting, Brearley's original insight – at least we believe it is Brearley's – is commonly being credited to Barrow.

[1] *Rowing and the Same-Sum Problem Have Their Moments*, American Journal of Physics, **78**, 728–732, 2010.

It is easy to see how such misattributions can occur. One can just hope that such mistakes are uncommon and are quickly corrected when they occur. Alas, mathematical life is very often not very fair. For various reasons, some innocent and some much less so, many mathematical ideas become attached to the wrong people.

As a striking example, take Pythagoras. He is the single mathematician we are most likely to remember from school, and he is of course immortalised by his *Pythagoras's Theorem.* Except, it seems fairly clear that the theorem is not actually Pythagoras's.

This grandest of all triangle theorems was known and used in ancient Mesopotamia, at least a thousand years before Pythagoras.[2] True, there is no evidence that the Babylonians knew how to *prove* the theorem. There is no evidence, however, that Pythagoras ever proved the theorem, either. Indeed, modern scholars strongly suspect that he did not.[3]

As another example, above is a Chinese illustration of "Pascal's" triangle, dated to 320 years before the birth of Blaise Pascal. And, in case the pattern isn't yet clear: *L'Hôpital's rule* was probably taught to L'Hôpital by Johann Bernoulli; *Cardano's formula* for solving cubic equations originated with Niccollò Tartaglia; Fibonacci wasn't the first to come up with his numbers; *Pell's equation* was so named by Leonhard Euler, even though Pell had only copied the equation from the letters of Pierre de Fermat; *Benford's Law* was discovered by Simon Newcomb, 60 years before Benford rediscovered it; the *Möbius strip* might as well be called Listing's strip; and on and on. All a bit depressing, really.

What is the moral of this story? Well, if your joy is purely in the mathematical discovery, then just don't worry, be happy. However, if your goal is eternal mathematical fame – and why not – then we suggest you give your theorem a catchy name, make it very easy to locate, and prepare an enticing media release.[4] That probably won't do the trick either, but at least you'll have a shot.

Puzzle to ponder

How can you make a perfect right angle using a loop with 12 equally spaced beads within the loop?

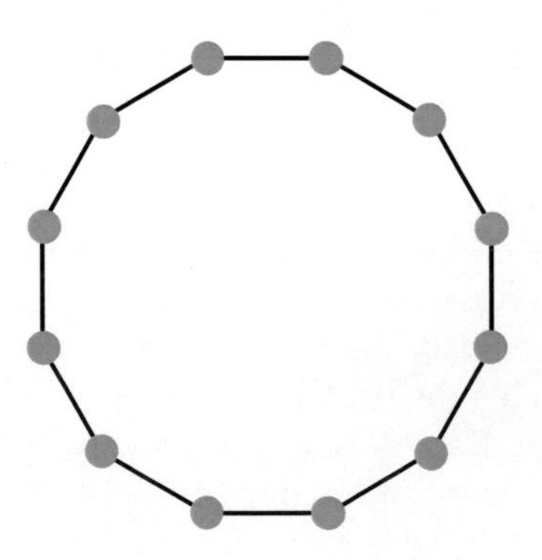

[2]See the next Chapter.

[3]See W. Burkert (translated by E. L. Minar Jr.) *Lore and Science in Ancient Pythagoreanism*, Harvard University, 1972.

[4]We're working on the Marty numbers. How are we going?

CHAPTER 59

Six of one, Babylonian the other

The Wonders of Mesopotamia is an excellent exhibition now on display at Melbourne Museum. The exhibition features fascinating Sumerian and Babylonian artifacts, many over 4000 years old. And, what is really excellent is that the very first artifact you meet is a mathematical gem.

"Babylon" has a special resonance for mathematicians. Referring both to the ancient Mesopotamian civilisation of Sumer and, more accurately, to the empires that followed, Babylon is generally considered to be the birthplace of mathematics. Moreover, later Babylonian mathematics was in many ways astonishingly advanced.

Of course, deciding when and where mathematics was born depends upon what we mean by "mathematics". Tally sticks have existed for at least 30,000 years, and it seems clear that counting in some form or another is much older. So, we might just declare the first mathematician to have been Og the Caveman[1]

Numbers only take on their full mathematical role, however, when we begin doing things with them. So it's perhaps more realistic to date mathematics to the beginning of measurement and arithmetic. And it seems that all that took off between 3000BC and 3500BC, coinciding with the emergence of cities and large scale agriculture. Of course for any such society to function, some arithmetic and geometry is essential: there are fields to measure, seasons to predict, products to weigh and to sell, and so on. So, if the Sumerians were the original mathematicians it was perhaps necessarily so; theirs was simply the first civilisation of sufficient size and complexity. This is where the questions begin, however.

A civilisation may well putter along with just enough mathematics to make do. That seems to have been true of the contemporaneous Egyptian kingdoms. It was most definitely true of the later, Roman Empire, which worked hard to squander

[1]Not his real name.

the wealth of Greek mathematics gifted to them. But, it was decidedly not true of the Babylonians.

By about 1600BC Babylonian mathematicians had made some stunning achievements: they knew about "Pythagoras's" Theorem, a good thousand years before Pythagoras was born;[2] they could solve essentially all quadratic equations, and many similar equations; they had excellent methods for approximating square roots; they had many formulas for areas and volumes of geometric figures. And, generally, it appears that the Babylonians developed a very strong sense of algebraic methods; their thinking was inventive and abstract.

True, there were also some relative clangers. Some of the Babylonian geometric formulas were badly inaccurate, and it seems that seldom if ever did they do better than 3 1/8 as an approximation for π. Nonetheless, all in all, Babylonian mathematics was genuinely brilliant, in some ways outclassing the much heralded Greek mathematics that followed.

So, how? How did Babylonian mathematics progress so far? No one knows for sure, but it seems likely that one peculiar aspect, dating back to early Sumerian times, had an enormous effect: Babylonians were really keen on the number 6.

That doesn't sound like much of an explanation. What's so thrilling about 6? And, how could a preference for one particular number lead to such general accomplishments?

More precisely, the key is the number 60, which is 6×10, and it all started with the early Sumerians' choice of 60 as a principle unit of measurement. The earliest Sumerian writing, around 3000 BC, contained the following numerals (all easily produced by poking a stylus into the clay tablets they used for writing):

1 10 60 600 3600 36,000

Then, general numbers were indicated by repeating the appropriate numerals, just as is done with Roman numerals. An example, one of the earliest recorded numbers appears below.

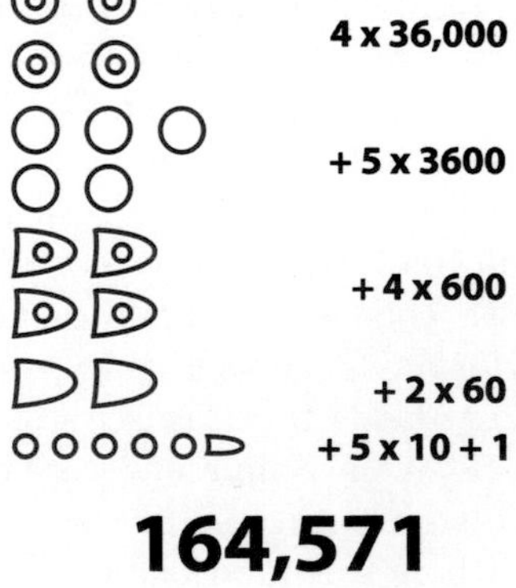

4 x 36,000

+ 5 x 3600

+ 4 x 600

+ 2 x 60

+ 5 x 10 + 1

164,571

[2]See the previous Chapter.

Given our own deci-centric society this choice of 60 may seem very strange, but it was, and is, natural and functional. Because 60 has so many factors, the Sumerians were able to handle many division problems with ease: fractions of one unit of measure were often whole amounts of another. It's exactly why we still divide an hour into 60 minutes, a circle into 360 degrees and so on.

It is not clear that the Sumerian use of 60 encouraged the next step, but a general ease of calculation couldn't have hurt. And the next step was huge: at some point, around 2000 BC, a very intelligent Babylonian had the revolutionary idea to introduce *positional notation.* This is the same system we use today, where numerals do double duty, or more. In the number 747, for example, the first 7 stands for seven hundreds, and the second 7 stands simply for the number seven. In this way, we can easily and economically write any whole number using only the ten numerals $0, 1, 2, \cdots, 9$.

The Babylonians used positional notation in exactly the same manner. So, they had a ones place, a 60s place, a $60 \times 60 = 3600$s place, and so on. Given the Babylonian method of repeating numerals, however, they got by with just two symbols:

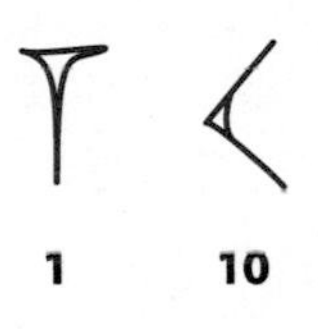

(The Babylonians had no symbol for 0 and no concept of 0 as a number, which created ambiguities in Babylonian positional notation. These ambiguities were clarified by spacing and context. Later, a placeholder symbol was introduced, which effectively played the role of 0.)

Now, for example, our huge Sumerian number above would be written in Babylonian positional notation as

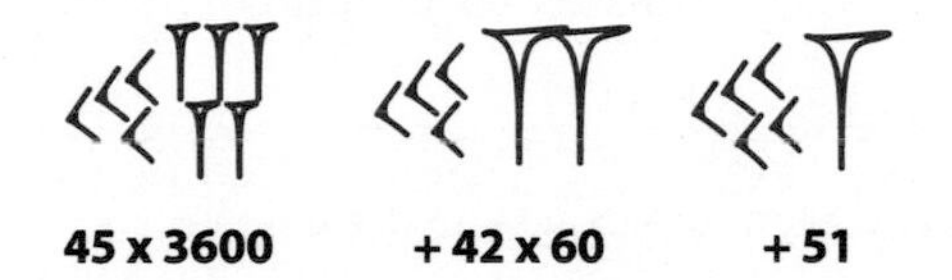

For the Babylonians to come up with positional notation was hugely impressive. It was not to be rediscovered for over two thousand years. But the Babylonians went further, and their next step was almost certainly facilitated by the use of 60 as a base.

We use positional notation to represent not only very big numbers but also very small ones: that is, we have the *decimal representation* of fractions. For example, we can write $1/4$ as 0.25, with 2 in the tenths place and 5 in the hundredths place. The Babylonians took the analogous step: they introduced a $1/60$ place, a $1/3600$ place and so on.

Moreover, due to 60's many factors, Babylonian *sexagesimal representations* were typically much simpler than our decimals. For example, since $1/4 = 15/60$ this fraction could be written by the Babylonians with just 15 in the $1/60$ place. An even more telling example is $1/3$, which didn't bother the Babylonians at all, while we have to resort to an infinite decimal.

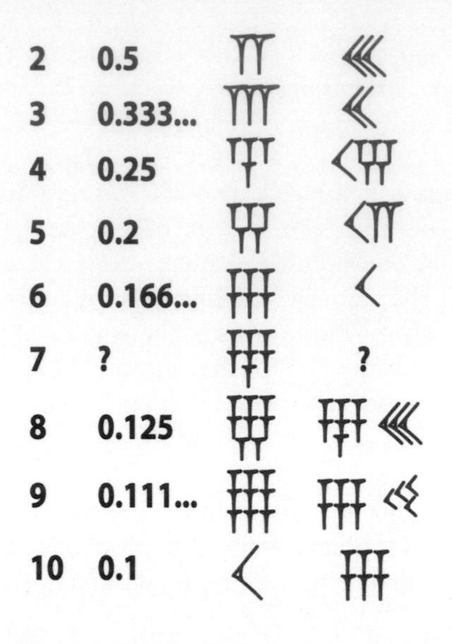

The above table of reciprocals illustrates the benefits of 60 as a base, and also that 60 was not a cure-all. The reciprocal 1/7 could only be expressed as an infinite repeating sexagesimal, just as we would write it as an infinite repeating decimal. Indeed, Babylonian tables of reciprocals often left out such problematic examples, though they did have methods for obtaining approximations to these fractions.

But the underlying truth remains, that the Babylonian numerical system made many arithmetic computations easy. That freed Babylonian minds to explore deeper ideas, and it made it cleaner and easier to explore those ideas. Whether the cause or not, the result was the creation of some very beautiful mathematics.

And that very first artefact in the Museum's exhibition? It is a clay tablet, one of these Babylonian tables of reciprocals. As we said, a mathematical gem.

Puzzles to ponder

What is π to one sexagesimal place?

What is the (infinite) Babylonian sexagesimal for 1/7?

CHAPTER 60

It's Chris Mass time

It's a very good year for mathematical anniversaries. We've already written for the centenaries of the birth of brilliant logician Alan Turing and the death of brilliant mathematician Henri Poincaré. This week we'll travel much further back in time.

First, we make quick mention of Ahmed Yusuf Ahmed, who died in 912, and was one of the Arabic mathematicians responsible for reviving the study of the long-forgotten Greek mathematics. Ahmed's works were translated into Latin, influencing early European mathematicians such as the famous Fibonacci. We don't know much more about Ahmed but it's not often that we have the opportunity to celebrate a 1200th anniversary. Well, give or take a year: it will come as no surprise that the precise date of Ahmed's death is uncertain. There is no such doubt in dating our second mathematician, however.

Christopher Clavius died 400 years ago, on February 6, 1612. How can we be so sure? One compelling reason is that Clavius is responsible for the calendar we still use today. From 45 BC to Clavius's time the Julian calendar had been in operation. Introduced by Julius Caesar, and with a careless mistake corrected by Emperor Augustus, the Julian calendar was based on a year of 365 days, with the inclusion of a "leap day" once every four years.

The Julian calendar was fairly accurate. It made for an average of 365 1/4 days per year, only a few minutes in error and still how we commonly think of the year's length. Over time, however, "only a few minutes" here and there can add up. By the 16th century, the accumulated errors had summed to ten days.

Not that many people cared: the Julian calendar worked just fine for everyday use. But being a week and a half off made the strict observance of religious holidays a trifle silly. In particular, the date of Easter depended upon the timing of the vernal equinox – the day on which the Sun moves from the Southern to the Northern Hemisphere – and the full moons, both of which had drifted from their "correct" dates.

It is no surprise that the call for calendar reform came from the ritual-conscious Catholic Church. After decades of discussion the reform finally eventuated in 1582, under the watch of Pope Gregory XIII. The result, with the Jesuit Clavius in charge of the mathematical calculations, was the Gregorian calendar. The key adjustment was to eliminate certain leap days, in those years divisible by 100 but not by 400.

We have written previously on the unsatisfiable desire for a mathematical exactness to Easter.[1] Due to the incommensurate lunar and solar cycles, no simple formula or method will work for long, except by religious decree. Moreover, even a complicated formula will eventually succumb to the inconstancy of the heavens. None of that stopped Clavius from using his new calendar to calculate the date of Easter millions of years into the future, long after his calculations would have any astronomical validity.

It was always going to be easy to nitpick Clavius's calendar. Françoise Viète, one of the founders of European algebra, devised his own calendar reform, along the way referring to Clavius as "a false mathematician ... and a false theologian". The dispute between them was typical, involving such compelling issues as the dating of full moons thousands of years in the future.

Objections to the new calendar in Protestant and Orthodox countries were very strong. These objections were egged on by papal arrogance, by the declaration that God would punish anyone who refused to accept the new calendar. It took Germany over a hundred years to adopt the calendar, and England even longer. Greece, the final nation to get on board, adopted the Gregorian calendar in 1923.

Nonetheless, the new calendar was also well received by many. The calendar was adopted by Catholic countries as planned, and Tycho Brahe and Johannes Kepler, the two greatest astronomers of the time, both supported the new calendar.

Though it did not affect the accuracy of Clavius's calendar, the 16th century was also a time of astronomical upheaval. Copernicus's heliocentric model of the solar system had been published in 1543 and the subsequent revolution was only a matter of time. Alas, Christopher Clavius backed the wrong horse, never accepting the Copernican theory.

Kepler's law of elliptical orbits was published in 1609 and Galileo made his telescopic discoveries in 1610, effectively dooming the geocentric theories. Clavius then died in 1612, too soon to properly evaluate these monumental events. He was destined to be the last great defender of a geocentric model of the solar system.

Almost immediately after Clavius's death, the Catholic Church firmed against the Copernican theory, opting to stick with Biblical literalism. There began the Church's infamous persecution of Galileo and its futile, foot-shooting denial of reality.

[1] See Chapter 54.

It is easy to make fun of Clavius, to see him as nothing more than a product of and promoter of the fundamentalist pseudoscience of his – and not only his – time. That would be wrong, and we're writing of Clavius not to bury him but to praise him.

In 16th century Rome, and elsewhere, mathematics was not held in high regard; it was considered too abstract to be of use. (Where have we heard that before?) More than anyone else of the time, Clavius dispelled that notion, raising mathematics to its deserved level in Jesuit teaching. He once wrote,

"Since ... the mathematical disciplines in fact require, delight in, and honor truth ... there can be no doubt that they must be conceded the first place among all the other sciences.

We couldn't have said it better. And Clavius meant it. Clavius believed not only that there was a real world out there to be examined, independent of scriptural guidance, but that the real world could be measured and the measurements possessed their own truth.

In 1611, not long before his death, Clavius and his colleagues met with Galileo. Clavius was (properly) cautious but acknowledged the reality of Galileo's discoveries. It seems he was also aware of the implications. For the 1611 edition of Clavius's textbook on astronomy, there was only time to include a brief description of Galileo's discoveries. But Clavius then commented on the discoveries:

"Since things are thus, astronomers ought to consider how the celestial orbs may be arranged in order to save these phenomena."

The precise meaning of this declaration has been hotly debated. No one knows how far Clavius would have been willing to go in accepting the Copernican theory. No one knows if he could have saved the Catholic Church – including the Jesuits – the centuries of ridicule that followed their trial of Galileo. Everything, however points to an openness, to an appreciation of and delight in truth, that would have stood Clavius in good stead.

Christopher Clavius indulged in religiously inspired silliness, and he ended up on the wrong side of history. But Christopher Clavius is a mathematical hero nonetheless.

Puzzle to ponder

Let's take a year to be 11 minutes and 14 seconds short of 365 1/4 days. What will be the accumulated error in the Gregorian calendar after 400 years?

CHAPTER 61

Clearing a logjam

What a difference a century makes. In 1914 the most prominent historians of mathematics gathered to celebrate the 300th anniversary of the birth of a great historical figure. Now, in 2014, barely a peep.

We speak, of course, of the *logarithm*, the creation of the Scottish baron, amateur mathematician and professional religious fanatic, John Napier. Napier labored twenty long years to produce his tables of logarithms but it was worth it; when the tables finally appeared in 1614 they took the mathematical world by storm.

But why was Napier's work so revolutionary? And why, after such grand celebrations in 1914, including the creation of a wonderful Tercentenary Memorial Volume,[1] is almost nothing planned for logarithms' 400th anniversary?

Answering the first question will require some explanation, and we'll begin by considering a simple example. Suppose we want to find the product of 512 and 256. Now, we could simply multiply the two numbers in the traditional manner

[1] *C. G. Knott, Napier tercentenary memorial volume*, Longmans, 1915.

but, well, we're lazy. (Insert Australian Curriculum joke here.) Luckily, we happen to have a table handy, listing the powers of 2:

number	power
1	2
2	4
3	8
4	16
5	32
6	64
7	128
8	256
9	512
10	1024
11	2048
12	4096
13	8192
14	16384
15	32768
16	65536
17	131072
18	262144
19	524288
20	1048576

Now 512 is 2 multiplied by itself 9 times, and 256 is 2 multiplied by itself 8 times. Aha! Since $9 + 8 = 17$, that means 512×256 must be 2 multiplied by itself 17 times. So, we locate the 17th entry in our table and there's our answer: 131072. Easy!

Of course it was, um, fortuitous that the numbers to be multiplied happened to appear in our table. But putting that aside for now, the technique is astonishingly clever. The effect is to transform a tedious multiplication problem into a simple addition, together with two applications of our table: first to locate the numbers in the left column corresponding to 512 and 256; then to transform backwards, from the sum 17 to our final answer in the right column.

And there, in a nutshell, you have logarithms. (The name, Greek for "number ratio", is due to Napier.) In the above context we would refer to the numbers 9 and 8 as logarithms base 2, and we would write $\log_2 512 = 9$ and $\log_2 256 = 8$. Then the heart of the above method is the (once) familiar log rule

$$\log_2 ab = \log_2 a + \log_2 b$$

Notice that the logarithm equation $\log_2 512 = 9$ means nothing more and nothing less than the power equation $2^9 = 512$. A sneaky way to relate the two equations is the very clever *sock rule*:

The above method of transforming multiplication into addition – and, similarly, division into subtraction – is so elegant, it's amazing that no one discovered it prior to Napier. Well, in fact, they did.

Tables similar to ours date back to the Babylonians and were known and studied in Europe centuries before Napier. In 1484 the French mathematician Nicolas Chuquet discovered what we now call (somewhat misleadingly) *index laws.* In particular, Chuquet understood the identity which we would now write as $2^c \times 2^d = 2^{c+d}$, exactly the property of powers we used above (with $c = 9$ and $d = 8$), and exactly the relationship expressed by the log rule above. Chuquet's work was not properly published until 1880, however 16th century German mathematicians were independently making the same discoveries. In particular, in 1550 Adam Ries noted that such a table could be used to calculate 512×256, precisely as we have above.

The trouble, of course, is that the gaps between the numbers in such tables are so great that the tables are generally useless, even for obtaining approximate answers. So, tricks such as Ries's, though exemplifying important mathematical identities, were not considered to have general application as a computational device. However, it is not as if the mathematicians and scientists of the time weren't on the lookout for some efficient method of multiplication and, even more so, division.

One remarkable, and remarkably titled, method employed by astronomers in the late 16th century was *prosthaphaeresis.* Similar to Ries's trick, prosthaphaeresis was a method of transforming the calculation of a product into that of a sum, doing so by way of converting the numbers to be multiplied into trigonometric quantities.[2] The method was ingenious and, since accurate tables of trigonometric functions were readily available (for navigation and other purposes), it was often simpler than multiplying directly. But, and this may be difficult to believe, multiplying by way of trigonometry was artificial and awkward, and was not hugely popular.

It seems likely that Napier knew of the method of prosthaphaeresis, and it is possible that the knowledge of that method spurred him on to find something better. Which he did. In effect, Napier simply fine-tuned Ries's trick to be much more accurate. To do so, all that was required was to make the gaps between the numbers in his table smaller, and there was an easy way to do that: just choose a base much closer to 1, so the table of powers would grow much more slowly. And that's it: Napier's tables are complicated in their definition, but it was roughly equivalent to his using a base of 1.0000001.

With historical hindsight Napier's idea was ingeniously simple. There were two excellent reasons, however, why it was anything but simple for Napier. First of all, the powerful notation (pun intended) a^b was not yet around, and the idea

[2]The heart of the method was the identity $\sin A \sin B = \frac{1}{2}\left[\cos(A-B) - \cos(A+B)\right]$, and similar identities, which convert a product of trig quantities into a sum.

of negative powers (as reciprocals) and fractional powers (as roots) were just then being sorted out. It seems likely that Napier wasn't aware of the index laws, even for positive whole numbers; Napier's description and explanation of logarithms was very geometric, almost physical, rather than algebraic. (The Swiss mathematician Jost Bürgi, who independently developed the theory of logarithms around the same time as Napier, was directly motivated by the index laws.) The second issue is that decimals, which obviously would have come in handy, were very new; Napier effectively avoided the use of decimals by scaling all his logarithms by a factor 10^7.

As a consequence of Napier's approach, his logarithms were awkwardly defined and awkward to use. In particular, they didn't obey the log rules in the simple form indicated above. But, it didn't take much to put things right.

British mathematician Henry Briggs immediately saw the value of Napier's work and contacted Napier. They met in 1615, by which time Napier had already decided that 10 was the most useful base for his logarithms – in effect employing fractional powers – and to do away with the scaling, employing decimals instead. The result was base 10 logarithms as we know them today. Napier died in 1617, but Briggs went on to produce a partial table of logarithms in 1624, which was then completed in 1628 by Danish publisher Adriaan Vlacq. Vlacq's tables became the source of almost all subsequent logarithm tables, and so were the constant companions of all mathematicians and scientists, and all schoolchildren, for the next 350 years.

Of course, once electronic calculators became cheap and readily available, log tables went the way of the dodo. And fair enough. Your Maths Masters are (just) old enough to have been taught with log tables and slide rules; they were very impressive, but they were still work, and it's not like they ever made arithmetic fun.

Nonetheless it is sad to see such a fundamentally beautiful idea disappear from general view. Sure, *natural* logarithms – logarithms to the base e – still play a pivotal role in calculus, but it's not the same; logarithms once played a pivotal role in *everything*.

So, it's not that we feel logarithms' 400th anniversary should have the same pomp and ceremony as its 300th. Of course we understand that their time as a computational tool has passed. Still, it is poignant, that mathematics that for over three centuries was so central to so many people, should just slip away, to be just another part of history.

Puzzle to ponder

Let's write Napier's logarithm function as nlog. Then, in modern terms, Napier's function is

$$\text{nlog}(a) = -10^7 \left(\log_e a - \log_e \left(10^7\right)\right) ,$$

where $e \approx 2.71828$ is the base for the *natural logarithm*.[3] How does the rule $\log(ab) = \log a + \log b$ translate into a rule for nlog?

[3] See Chapter 7.

CHAPTER 62

Squares, triangles and other labor-saving devices

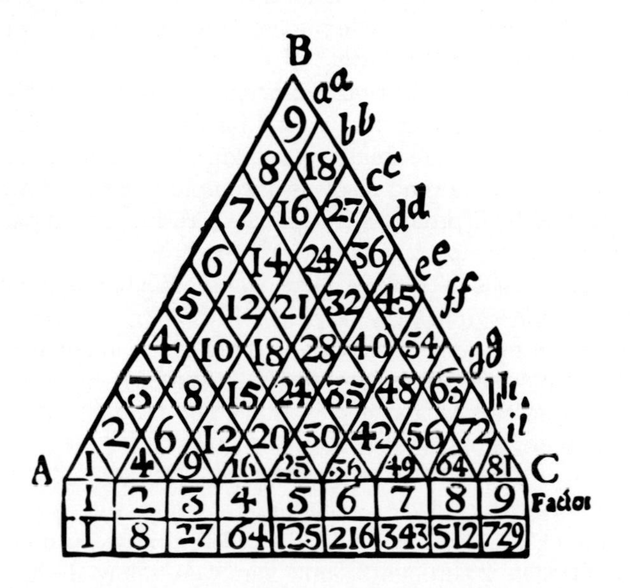

Recently we wrote about logarithms, which are celebrating their 400th anniversary this year.[1] The magic of logarithms is that they can transform an annoying multiplication problem into a much easier addition, at the small price of a couple look-ups in tables. The introduction of logarithms was revolutionary, but it was also a single episode in a very long history. Mathematicians have *always* been lazy, have always sought to avoid the hard work of multiplying.

How else could we avoid multiplying? Yes, yes, we could use a calculator. But let's concentrate upon historical approaches and leave the discussion of zombie techniques for another day. (Insert Australian Curriculum joke here.)

One remarkable method, which we have mentioned previously, is *prosthaphaersis*.[2] Discovered in the 16th century this ingenious but clumsy technique applies trigonometric formulas to transform multiplication into addition. Logarithms arrived soon after and prosthaphaersis fell into disuse. But, in any case, there *had* to be something easier than exotic trig tricks.

Simplest of all, at least for whole numbers, is to have multiplication tables on hand, or in head. For example, if we know our 12 times table then there is no need to calculate 7×12: we simply know that the answer is 84. So, it's a good thing we all know our 12 times table, isn't it? (Insert Australian Curriculum joke here.)

---

[1] See the previous Chapter.

[2] See the previous Chapter.

It is impractical to memorize multiplication tables of any significant size but we could still hope to have access to written tables. The Babylonians employed multiplication tables over 4000 years ago and such tables have pretty much always been around. (Insert ... nah, it's getting too depressing.)[3]

Multiplication tables typically include a very restricted range of multipliers and are intended to aid calculations rather than replace them. But why not simply have a table with a large range of multipliers?

These tables exist. In 1610, the German statesman Herwart von Hohenberg published a 1000×1000 multiplication table; this table has been reconstructed with an historical introduction by mathematics historian Denis Roegel.[4] In 1820 the German mathematician A. L. Crelle published a table of the same range, which became very popular and is still in print.[5]

The problem is that multiplication tables quickly become prohibitively large. Hohenberg's table(s) was 999 pages long, one page for each multiplier from 2 to 1000; Crelle took some clever shortcuts, as explained by Roegel, making the presentation more elegant, but the end result is of comparable size. However it's done, we have about 1000×1000 products, meaning we're in the ballpark of a million entries. A table extending up to 10,000 would require around 100 million entries which, at an optimistic 1000 entries per page, would make for a 100,000 page book.

We cannot hope for significantly larger tables, but notice that Crelle's tables can also be applied to multipliers that are not explicitly included. For example, his tables can be employed to calculate 38.4×2.56 simply by shifting a few decimal points back and forth. As another example, the product $23 \times 36,145$ can be calculated as the sum of 23×145 and $23 \times 36,000$.

Such arithmetic manipulations can greatly extend the application of multiplication tables and Crelle included instructions for doing so. It also suggests the usefulness of tables of other forms. In 1836 Crelle published product tables up to 10,000,000 for multipliers 2 through 9, and there were many similar tables around at the time.[6]

So it turns out multiplication tables such as Crelle's are much more useful than it might first appear. Still, they make for a very big book. Luckily, there turns out to be some very clever alternatives.

Given two numbers A and B, a little algebra shows that

$$A \times B = \frac{1}{4}\left[(A+B)^2 - (A-B)^2\right] .$$

---

[3]Yes, we worry about this. But we worry about this because it's way worrying. We're firmly convinced that unless and until it is accepted that primary students should know their multiplication tables, up to 12, *by heart*, then there's no point in discussing anything else about mathematics education.

[4]Roegel and others have reproduced and reconstructed an astonishing number of such tables, including some of the tables to which we refer, below. At the time of writing, these tables are available online.

[5]*Dr. A. L. Crelle's Calculating Tables Giving The Products Of Every Two Numbers From One To One Thousand And Their Application To The Multiplication And Division Of All Numbers Above One Thousand*, Palala Press, 2015.

[6]In 1873, the British Association for the Advancement of Science published a remarkable historical survey of mathematical tables.

This pretty formula has a striking consequence: if we can calculate squares then it is not much extra work to calculate general products. For example, to calculate 7×12 we can set $A = 12$ and $B = 7$. Then $(12+7)^2$ equals 361, $(12-7)^2$ equals 25, and the difference is $361 - 25 = 336$. A quarter of 336 is 84, which is the product we're after.

Though very clever, this method may not seem very helpful: computing squares is hardly more fun than general multiplication. If, however, there is a table of squares handy then we'd only have to perform a few additions and subtractions, and a final, annoying division by 4. And the critical point is that tables of squares can be much, much smaller than multiplication tables. A table of squares up to 1000, for example, would have just a thousand entries rather than the million-ish for Crelle's multiplication table.

Tables of squares have been around since Babylonian times but it is possibly not until the late 17th century that they were used for multiplication. (It is commonly claimed that the Babylonians did so but these claims appear to be somewhat speculative.) In 1690 the Dutch mathematician Johann-Hiob Ludolf published a table of squares up to 100,000 and specifically noted the application of his table to multiplication.

One drawback with the squares method is the annoying division by 4. But, if we rewrite our equation as

$$A \times B = \frac{(A+B)^2}{4} - \frac{(A-B)^2}{4}.$$

then it is clear what we actually need is a table of "quarter-squares", which would circumvent the final division. In 1817 a table of quarter-squares up to 10,000 was created for this very purpose, and in 1887 a table up to 200,000 appeared.

There is, however, a second, subtle drawback with the squares – or quarter-squares – method: the method can only be applied if our squares table extends as far as the sum $A + B$. That's exactly the reason the 1887 table extended to 200,000, so that the product of any two five-digit numbers could be calculated. Slightly more complicated versions of the square formula avoid this problem, but at a significant cost to convenience. But, there is another clever device to avoid this problem entirely: *triangular numbers.*

Triangular numbers can be obtained by counting the number of dots placed regularly in an equilateral triangle. As pictured below, the sequence of triangular numbers is then 1, 3, 6, 10, and so on.

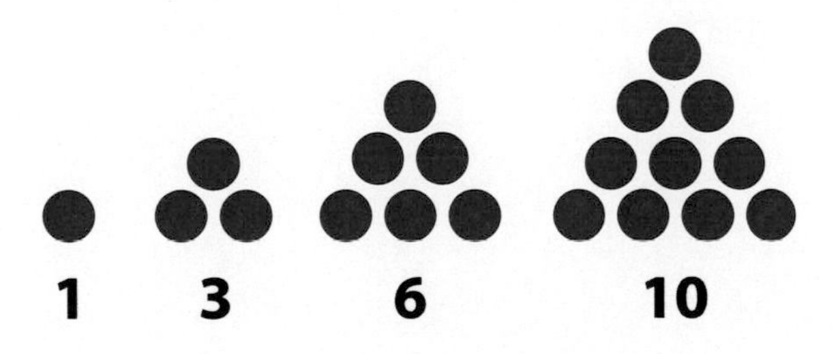

Writing $T(N)$ for the Nth triangular number, it is not all that difficult to show that $T(N)$ is equal to $N \times (N+1)/2$. Triangular numbers are natural and interesting, and a large table of them was published in 1762. In 1889, the English mathematician James Glaisher noted that triangular numbers could be used to calculate products, by using a beautiful and simple formula:

$$A \times B = T(A-1) + T(B) - T(A-B-1)\,.$$

For example, calculating 7×12 once again, we can set $A = 12$ and $B = 7$. Then $T(11) = 66$, $T(7) = 28$ and $T(4) = 10$. We then calculate 7×12 to be $66 + 28 - 10$, which comes to 84. Easy.

For Glaisher's method, the table needn't extend any further than A and B, and it was not much slower than the method of quarter-squares. Nonetheless, the method never caught on. Indeed, quarter-squares never really caught on; although quarter-square tables regularly appeared up until the 1950s, the method never gained the general acceptance that one might expect.

So, what did people actually do, at least what did they do before the triumph of the zombie calculators? It seems that Crelle's table, with perhaps a little arithmetic, were considered quick and easy for many calculations. Otherwise, logarithms were favored, because of the flexibility of their application.

The quarter-square and triangular methods of multiplication are very clever and very pretty, and it's a shame they were never more appreciated. But sometimes that's the mathematical way of things, that the prettiest is simply not the best.

Puzzle to ponder

Prove the formula for triangular numbers:

$$A \times B = T(A-1) + T(B) - T(A-B-1)\,.$$

CHAPTER 63

The doodle, the witch and Maria

It's a fun little moment each day, pondering the new Google doodle, seeing which famous artist or scientist, or thing, Google has elected to honor. And on May 16 it was mathematics' turn, with the celebration of the 296th birthday of Maria Gaetana Agnesi.

Alright, hands up if you've never heard of Maria Agnesi. Come on, hands up high. Hmm. Pretty much everyone.

So who was Maria Agnesi? Why has Google decided to honor her? Was she a witch?

The Italian Maria Agnesi was a truly remarkable individual and most definitely not a witch. She first became famous among her contemporaries as a child prodigy, mastering seven languages by the age of 11, debating philosophers and the like. Maria was also deeply religious and in her teens wished to become a nun. She disliked performing, much to the annoyance of her social climbing father, and sought a quiet and reflective existence. After her father's death Maria was fully free to follow her heart; she dedicated the rest of her long life and all of her significant wealth to helping the poor, finally dying penniless in one of the poorhouses she had directed. Maria Agnesi was a model of devotion and charity (and an example that a current day Christian or two might do well to ponder).

But how about the math? Here, things are less flattering to Maria, and a little controversial. The most informative article on Maria Agnesi's mathematical legacy is a remarkably forthright survey by mathematician and historian Clifford Truesdell.[1]

Today Maria Agnesi's name is mainly known through the *witch of Agnesi*, the popular name for the mathematical curve running through Google's doodle. The animation in the doodle elegantly illustrates a geometric method of constructing the curve, beginning with a horizontal (tangential) line perched atop a circle.

[1] *Maria Gaetana Agnesi*, Archive for History of Exact Sciences, **40**, 113–142, 1989.

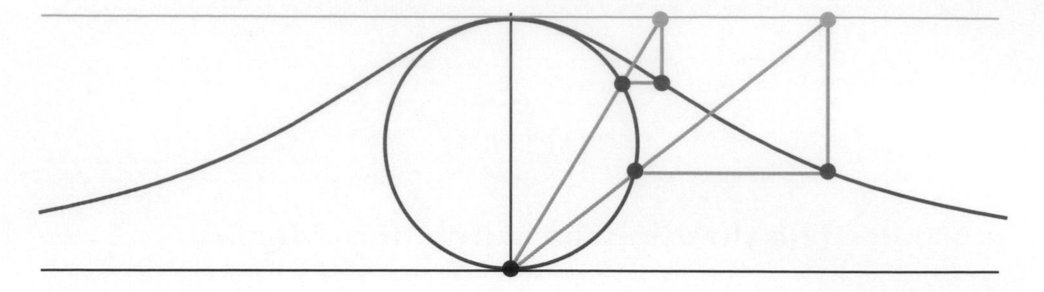

As pictured above, for each point on the green line we draw a straight line to the South pole of the circle. This line will intersect the circle in a second, purple point. Then, the blue point completing the right-angled triangle will be a point on the desired curve.

There is nothing particularly memorable, and certainly nothing particularly witchlike, about this curve. If the circle is chosen to have radius $1/2$ then the resulting curve is the graph of $y = 1/(1 + x^2)$, familiar to most senior mathematics students but probably as "just another example". Moreover, Agnesi didn't discover the curve, or name it, or declare any special interest in it. So why all the fuss over a rather ordinary curve and wherefrom its weird name?

In 1748, Agnesi published a calculus textbook which featured the above curve, along with hundreds of others. She referred to the curve by its familiar Italian name, *la versiera*, meaning (more or less) to curve in every direction. About a decade later the English mathematician John Colson translated Agnesi's textbook into English and made the critical, memorable mistake: Colson confused la versiera with *l'avversiera*, which roughly translates as "the witch". The meaningless but catchy title caught on and the curve's fate was sealed.

Misnamed curve aside, Agnesi's textbook seems to have been somewhat influential. It was well-received in Italy and was later translated into French and, as mentioned above, English.[2] It is possible, however, that the book's influence has been overstated.

Truesdell attributes the translating of Agnesi's text not to its great merits, but to the poorly judged occupation of some of mathematics' lesser lights. Truesdell doesn't mince words, referring to "the tenacious, continuing grip of ignorant, bloated triflers on the mathematical scene". John Colson, in particular, he dismisses as "one of the feeblest of [Sir Isaac] Newton's feeble successors". (Your Maths Masters really wish we were permitted to write in Truesdell's style.)

Not that Truesdell doesn't have good things to say about Agnesi's work. He agrees with most everyone in describing her textbook as very carefully written and a model of clarity, at a time when few if any introductory texts existed. (Truesdell describes one text of the time as requiring a strong stomach, and another as "so dense, torve and ugsome as scarcely to have been read through by anyone but its author".) It was more the case of the wrong topics and the wrong style at the wrong time. Just as Agnesi's text appeared so did the great Leonhard Euler's monumental *Introductio in Analysin Infinitorum*, which was to revolutionize mathematics and the learning of mathematics.

---

[2]At the time of writing, the English translation of Agnesi's textbook can be found online.

In any case, after Agnesi's textbook appeared Pope Benedict XIV offered her a professorship at the University of Bologna. So she must have been a very good mathematician, right? Well, no.

Maria Agnesi definitely understood the mathematics in her textbook, but it is clear from her correspondence with contemporary mathematicians that she was really just a beginner trying to make sense of the subject as she went along. None of her mathematics was original, and she never pretended otherwise. It seems that her purpose in writing the book was simply to learn mathematics and, as for so many of her endeavors, to help others.

So why all the attention? Both then and now Maria Agnesi has served as a symbol. She was a kind and deeply intelligent woman, who shone at a time when women were seldom even permitted to compete. Maria Agnesi was so genuinely admirable, one is easily tempted to claim her successes to be greater than they were.

That appears to be the case with the evaluation of Maria Agnesi's mathematics. Writing a mathematics text, even a very good one, does not make a person a good mathematician. The reality is that Maria Agnesi was never more than a very thoughtful, very diligent beginner.

In brief, if Maria had instead been Marty there never would have been an Agnesi Google doodle. But we're very glad that there was, that we were encouraged to learn some of the truth about this truly remarkable person.

Puzzle to ponder

Starting with a circle of diameter 1, show that Maria's witch has equation $y = 1/(1+x^2)$.

CHAPTER 64

Christian Goldbach's magic sum

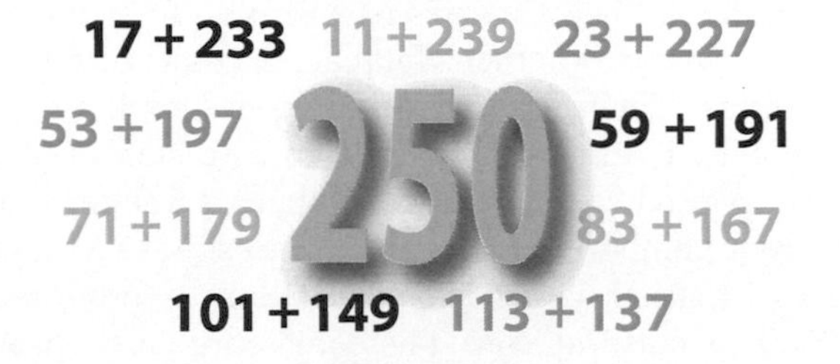

This week we're pleased to present *The Age*'s 250th Maths Masters column. Since 2007 we've covered plenty of territory and your Maths Masters are frequently asked whether we're running out of ideas. The answer is a resounding "Yes": next week's topic is Snakes and Ladders and that's about it.[1] Then we'll just keep banging on about Tony Abbott and Australian mathematics education until our readers run us out of town.[2]

We still have one meaty mathematical topic left, however. This week we acknowledge another, much more significant 250 milestone, the 250th anniversary of the death of German mathematician Christian Goldbach.

Goldbach may not have been one of the great mathematicians but he was no slouch, and he had the good sense to correspond with a number of the greats: Gottfried Leibniz, lots of Bernoullis and, most famously, the mega-great Leonhard Euler. It is largely for this correspondence that Goldbach is now remembered, in particular for a 1742 letter to Euler on natural numbers – positive whole numbers – and what is now famously known as *Goldbach's conjecture.*

Goldbach's conjecture is tantalizing because it is so easy to understand. It is the simple statement that *any even natural number greater than 2 can be written as the sum of two prime numbers.* So, for example $4 = 2+2$, $6 = 3+3$, $8 = 3+5$ and so on, hopefully to infinity. (Actually, Goldbach, along with many of his contemporaries, considered 1 to be a prime number. Consequently he would have regarded that $2 = 1 + 1$ can also be written as the sum of two "primes".)

Is Goldbach's conjecture true? No one knows, although Euler definitely believed so. He wrote that the conjecture was "an entirely certain theorem in spite of that I am not able to demonstrate it". Mathematicians today are just as confident, but why? After all, 270 years is a long time to be waiting for a proof.

Suppose we plan to write each of the even numbers below 1,000,000 as a sum of two prime numbers. The first question is how many primes do we have at our disposal, and the answer is over 70,000. But that means if we consider the different

[1] See Chapter 45.

[2] In the end, we crawled our way to 255 columns.

ways of summing two of these primes then we have over $70,000 \times 70,000/2$, over two *billion* pairs. That's a *lot* of pairs, enough for each even number to be written in 4000 different ways.

That's by no means a proof of Goldbach's conjecture, even for our limited range: some even numbers may simply be unlucky and be missed altogether. But we do know that any unlucky even number must be huge: Goldbach's conjecture has been checked by computer up to 4,000,000,000,000,000,000. Moreover, similar statistical estimates for the frequency of primes all the way to infinity provide strong circumstantial evidence that Goldbach's conjecture is true. As for Euler, he didn't have nearly as much data or as much knowledge as modern mathematicians, but Euler knew a lot, and he had an unsurpassed intuition for numbers.

It is impossible to predict whether Goldbach's conjecture will be proved next year or not for centuries. There has been some remarkable progress, however. It is known that every even number can be written as a sum of at most six primes. And just last year, a related conjecture of Goldbach's was proved; this conjecture implies that every even number can be written as a sum of four primes.[3]

Anyway, while we patiently wait for the definitive proof we can appreciate Goldbach's conjecture as cultural icon. The conjecture has appeared in *Futurama*, it was a motive for murder in a TV mystery,[4] it played a feature role in a Spanish movie,[5] Jimmy Stewart pondered it,[6] and it was the central theme of a popular novel.[7]

There's no end to the amount we could write about Goldbach's conjecture, but we want to turn our attention to another gift of Goldbach's and Euler's correspondence. In a (now lost) letter to Euler, Goldbach discussed a remarkable infinite sum. This sum, which we'll call GOLD, begins

$$\text{GOLD} = 1/3 + 1/7 + 1/8 + 1/15 + 1/24 + 1/26 + 1/31 + 1/35 + 1/48 + \cdots.$$

Hmm. So, each term is the reciprocal of a natural number, but what's the pattern? The clue is to add 1 to each denominator, producing the numbers 4, 8, 9, 16, 25, 27, 32, 36, 49. So, every denominator is 1 less than a power: $2^2, 2^3, 3^2, 4^2$, and so on. Goldbach's infinite sum includes the fractions resulting from all such powers, without repeating any fraction.

But how can we calculate such a sum, and why would we think it sums to anything interesting? Well, the only real answer to the second question is "Think like Euler". At the beginning of one of his amazing papers, Euler presents a remarkable method of calculating *Gold,*

Let's start with the total sum, of all the reciprocals from 1/2 on:

$$\text{TOTAL} = 1/2 + 1/3 + 1/4 + 1/5 + 1/6 + \cdots.$$

[3] H. A. Helfgott, *The Ternary Goldbach Conjecture is True.* Helfgott's paper has been accepted for publication by Annals of Mathematics Studies, and at the time of writing is available online. The wonderful comic *xkcd* (1310) also provides a convenient and very funny summary of "Goldbach conjectures".

[4] The pilot episode of *Inspector Lewis*, 2006.

[5] *Fermat's Room*, 2007.

[6] *No Highway in the Sky*, 1951.

[7] *Uncle Petros and Goldbach's Conjecture*, by Apostolos Doxiadis.

We first want to get a sense of the value of TOTAL. The scary but brilliant idea is to regroup the sum into infinitely many infinite sums, as follows:

$$\begin{aligned}
\text{TOTAL} \quad = \quad & \left(\frac{1}{2}+\frac{1}{4}+\frac{1}{8}+\cdots\right) \\
+ & \left(\frac{1}{3}+\frac{1}{9}+\frac{1}{27}+\cdots\right) \\
+ & \left(\frac{1}{5}+\frac{1}{25}+\frac{1}{125}+\cdots\right) \\
+ & \left(\frac{1}{6}+\frac{1}{36}+\frac{1}{216}+\cdots\right) \\
+ & \left(\frac{1}{7}+\frac{1}{49}+\frac{1}{343}+\cdots\right) \\
+ & \left(\frac{1}{10}+\frac{1}{100}+\frac{1}{1000}+\cdots\right) \\
+ & \cdots .
\end{aligned}$$

So, each bracket begins with the reciprocal of a non-power, and the subsequent terms are powers of that first term. (It takes a little thought to see that each fraction in *Total* still appears exactly once.) And the point is, each individual bracket can be easily summed.

The first bracket, $1/2 + 1/4 + 1/8 + \cdots$, is probably most familiar, an example of a *geometric series*. It is not difficult to believe that the bracket sums to 1, and this can easily be proved with a little pretty trickery.[8]

Indeed, every bracket is a geometric series, and all the sums can be calculated by the same trick. The sums are $1, 1/2, 1/4, 1/5, 1/6, 1/9$, and so on. So,

$$\text{TOTAL} = 1 + \frac{1}{2} + \frac{1}{4} + \frac{1}{5} + \frac{1}{6} + \frac{1}{9} + \cdots .$$

That is, TOTAL sums to 1 more than all the reciprocals that do *not* appear in the sum GOLD. So, adding GOLD to both sides, we get

$$\text{TOTAL} + \text{GOLD} = 1 + \text{TOTAL} .$$

Finally, subtracting *Total* from both sides, we conclude that $Gold = 1$. Amazing.

Theorema I.

Huius ſeriei in infinitum continuatae

$$\tfrac{1}{3}+\tfrac{1}{7}+\tfrac{1}{8}+\tfrac{1}{15}+\tfrac{1}{24}+\tfrac{1}{26}+\tfrac{1}{31}+\tfrac{1}{35}+\text{etc.}$$

cuius denominatores vnitate aucti dant omnes numeros, qui ſunt poteſtates vel ſecundi vel altioris cuiuſuis ordinis numerorum integrorum, cuiusque adeo terminus quisque exprimitur hac formula $\frac{1}{m^n - 1}$, *denotantibus* m *et* n *numeros integros vnitate maiores; huius ſeriei autem ſumma eſt* $= 1$.

[8]See Chapter 24.

Even more amazing, the proof as Euler presented it is complete nonsense. And, morely more amazing, though Euler's proof is nonsense it is truly true that Goldbach's sum is equal to 1. This result is now known as the *Euler-Goldbach theorem.*

What's wrong with Euler's proof? The problem is that Total sums to infinity,[9] which means that Euler has simply proved that

$$\mathrm{GOLD} + \infty = 1 + \infty\,.$$

Euler may then blithely subtract infinity from both sides, but it's the kind of mathematical move that would normally attract a big red cross on your school homework.

Euler's calculation is in the same spirit as his, let's say, "association" of the sum $1 + 2 + 3 + 4 + \cdots$ with the fraction $-1/12$.[10] The difference is, whereas Euler's purported answer for the sum of natural numbers is fundamentally not true, his calculation of GOLD above can be performed more carefully, in a completely justifiable way.[11]

Euler is famous for such magic tricks and Goldbach tends to be presented as the pedestrian one, merely as Euler's sounding board. Undoubtedly Euler was the infinitely greater mathematician, but Goldbach shouldn't be underestimated. In fact, it seems that the above calculation of GOLD is due to the "Celebrated Master" Goldbach, not Euler; at least, Euler claims as much in his paper.

Perhaps Goldbach was such a valued muse because he was also able to think magically, in the spirit of Euler himself.

Puzzle to ponder

Prove Goldbach's conjecture.[12]

How many even numbers below 100 can be written as the sum of two primes in exactly one way?

---

[9]See Chapter 27.

[10]We are not aware of any evidence that Euler wrote that $1 + 2 + 3 + 4 + \cdots$ *equals* $-1/12$, or that he believed the equation to be true in some sense. This equality was argued recently, however, in a very popular and very silly video.

[11]L. Bibiloni, P. Viader and J. Paradís, *On a series of Goldbach and Euler*, American Mathematical Monthly, **113**, 206–220, 2004.

[12]What? It's our last chapter. Can't we have a little fun?

Appendix: Solutions to the puzzles

1. Cordial math.

An easy example is where both numerators are 0, and then we have $0/b + 0/d = (0+0)/(b+d)$. At least this is an example if $b + d \neq 0$, which might happen if one of the denominators is negative. And, other similar things can happen if we allow negative numbers.

If all the numbers are positive, however, it follows from our cordial theorem that $(a+b)/(c+d)$ cannot possibly equal $a/b+c/d$. Why? Well, the cordial theorem tells us c/d is already bigger than $(a+b)/(c+d)$, and adding a/b only makes it bigger.

To prove the cordial theorem, we're assuming $a/b < c/d$, and so cross-multiplying tells us that $ad < cb$. Now, we want to show $a/b < (a+b)/(c+d)$, and after cross-multiplying, that's the same as showing

$$a(c+d) < (a+b)c\,.$$

Expanding, notice that there's an ac on both sides of that inequality, and what's left to show is $ad < cb$, which is what we're assuming. Proving $(a+b)/(c+d) < c/d$ can be done the same way.

A final teaser: where did our proof use the fact that a, b, c and d are positive?

2. Uncovering base motives.

Our alien is staring at π.

We know that in base three, $1 \times 1 = 1$ and $2 \times 2 = 11$. And this promises us that, ignoring the final zeroes, any square number written in base three must end in a 1. To see this, suppose we have a number A written in base three: $A = abc \cdots y00$, with y the final non-zero digit. There are only two possibilities, either $y = 1$ or $y = 2$, and in either case the final digit of y^2 is a 1. And *that* shows – just by doing the normal longhand arithmetic, as on our blackboard – the square A^2 must end as $cde \cdots 10000$. That is, ignoring the final four zeroes, A^2 ends in a 1.

To prove that $\sqrt{3}$ is irrational, we can think of writing numbers in base five. In this case, ignoring the zeroes, a square A^2 must end in either a 1 or a 4. And then, $3B^2$ must end in either a 3 or a 2. Again, the final digits cannot match, showing we cannot solve $A^2 = 3B^2$.

So now the question: given any non-square N, can we always find a base to prove that $\sqrt{N}$ is irrational. The answer is "yes". That is, for any non-square N, there is a really easy base-based proof that $\sqrt{N}$ is irrational. *However*, to *prove* that there exists such a base for every N, to *prove* that there is this easy base-based proof, is incredibly difficult. It involves a mysterious creature called *quadratic reciprocity*. Burkard has a *Mathologer* video devoted to this beautiful mathematical monster.

3. Marching in squares.

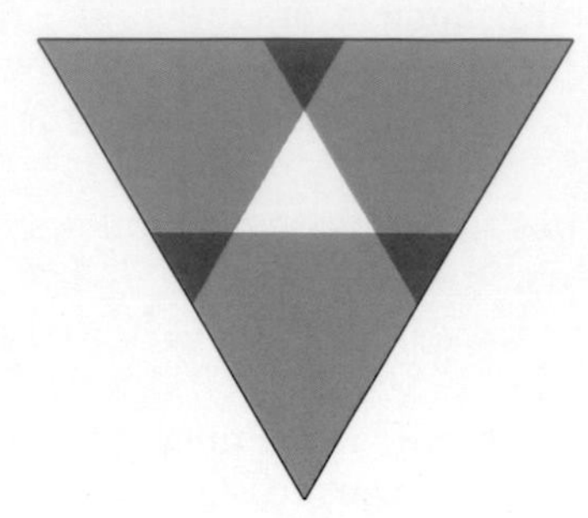

4. A very strange set of blocks.

Let S stand for the whole sum:

$$S = \frac{1}{2} + \frac{1}{4} + \frac{1}{8} + \frac{1}{16} + \cdots.$$

Then, multiplying both sides of the equation by 2, we get

$$2S = 1 + \frac{1}{2} + \frac{1}{4} + \frac{1}{8} + \cdots.$$

Now, subtract the first equation from the second. The left hand side is easy: $2S - S = S$. What about the right hand side? Notice that everything will cancel, the $1/2 - 1/2$ and so on, *except* for that starting 1 in the second equation. So, on the left we get S, and on the right we get 1. Left equals right, and so $S = 1$. Done.

The exact same style of argument can be used to prove $0.999\cdots = 1$.

5. Parabolic production line.

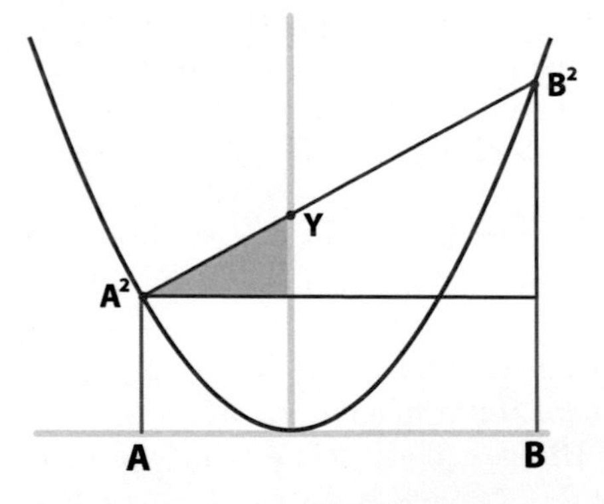

Calculating the ratio vertical/horizontal of the two similar triangles we find,

$$\frac{Y - A^2}{A} = \frac{B^2 - A^2}{A + B}.$$

Cancelling on the right and then solving easily gives $Y = AB$.

6. The magic of the imaginary.

Since $\phi = \frac{1+\sqrt{5}}{2}$, we can easily calculate

$$\phi^2 = \left(\frac{1+\sqrt{5}}{2}\right)^2 = \frac{1}{4}\left(1 + 2\sqrt{5} + (\sqrt{5})^2\right).$$

Now, since $(\sqrt{5})^2 = 5$, we end up with $\phi^2 = \frac{1}{2}(3+\sqrt{5})$. It is very easy to show that $1 + \phi$ sums to the same thing.

Similarly, and just using $(\sqrt{-2})^2 = -2$, we can calculate

$$\mu^2 = \left(1 + \sqrt{-2}\right)^2 = 1 + 2\sqrt{-2} + (\sqrt{-2})^2 = -1 + 2\sqrt{-2}\,.$$

It is easy to see that $2\mu - 3$ gives the same answer.

7. There's no e in Euler.

We need to investigate the quantity $(1 + 1/N)^N$, with N standing for our various compounding banks. You can just expand this quantity, and pick the right pieces out of the mess, but much easier is to use *Pascal's triangle*: see Chapter 58. This mathematical gem tells you what you get if you expand the quantity $(1 + x)^N$. In particular, the *second* number in each row of Pascal's triangle just goes $1, 2, 3, 4, \cdots$. So, the second number in the Nth row of the triangle is just N. This tells you that

$$(1 + x)^N = 1 + Nx + \text{Other Stuff}.$$

Applying this formula to our banks, we get

$$\left(1 + \frac{1}{N}\right)^N = 1 + \left(N \times \frac{1}{N}\right) + \text{Other Stuff} = 2 + \text{Other Stuff}.$$

Since the Other Stuff is all positive, this shows that each of our compounding banks returns at least \$2. It follows that Bank Infinity does as well.

One can similarly show $c \leqslant 3$, but it is trickier: one has to look at *every* term in the expansion of $(1 + x)^N$. For that one needs the *binomial formula*, which is a *formula* for all of the numbers in Pascal's triangle, beyond just a (very nice) method for churning them out. Proving $e \leqslant 3$ is a nice (but not easy) problem for those who know the binomial formula, but it is too involved to present here.

8. What's the best way to lace your shoes?

These are all the lacings up to symmetries. There are 6 very useful lacings, corresponding to the 4 in the box, and 42 lacings overall.

9. Ringing the changes.

For three bells, once the first change has been chosen, all subsequent changes are forced. This means that there are just two complete ringing sequences for three bells. The second sequence is the first sequence rung in reverse.

123	123
213	132
231	312
321	321
312	231
132	213
123	123

10. Triangle surfer dude.

If the common length of the sides of a regular pentagon is s, then the area of the pentagon is equal to the sum of the five colored triangles in the following diagram.

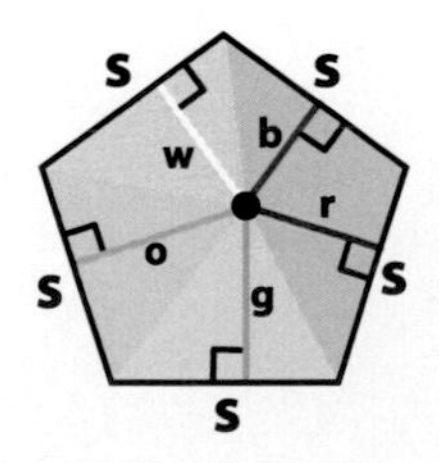

And so the area of the regular pentagon is

$$\text{Area} = \frac{s \cdot g}{2} + \frac{s \cdot r}{2} + \frac{s \cdot b}{2} + \frac{s \cdot w}{2} + \frac{s \cdot o}{2} = \frac{s}{2}(g + r + b + w + o)$$

This implies that the sum on the right, the sum of the distances we are interested in, is constant.

Next, placing this regular pentagon inside an equiangular pentagon as in the following diagram shows at a glance that our equiangular pentagon also has the equal-distance property we are interested in.

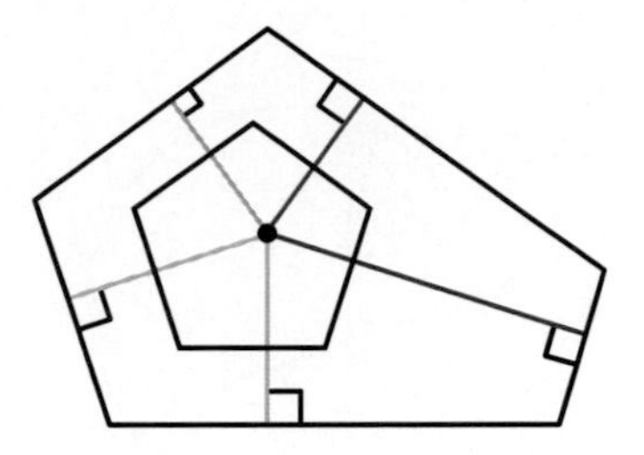

11. We have it pegged.

Using exactly the same argument as for the square board, we can convince ourselves that the first puzzle board is also unsolvable.

		L	LR	R		
	L	LR	R	L	LR	
L	LR	R	L	LR	R	L
LR	R	L	LR	R	L	LR
R	L	LR	R	L	LR	R
	LR	R	L	LR	R	
		L	LR	R		

All that changes is that instead of 16 of each type of label we now have 12 of each type.

The second game is possible. Give it a try.

12. Shadowlands.

As you can see in the following diagram, if the radius of the sphere is R then the radius of the shadow circle of the equator is $2R$. This means the area of this shadow is $4\pi R^2$, which also happens to be the area of the sphere.

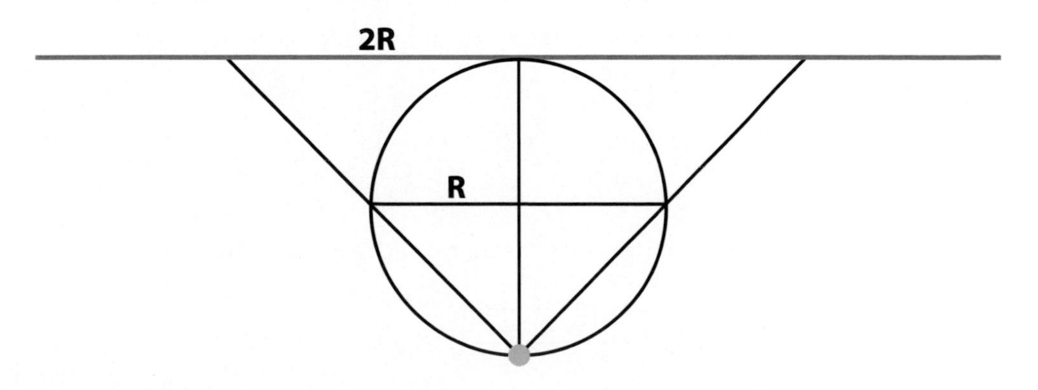

13. Picture perfect.

Let the point X be the intersection of the angle bisector and the circle through A, B, and C.

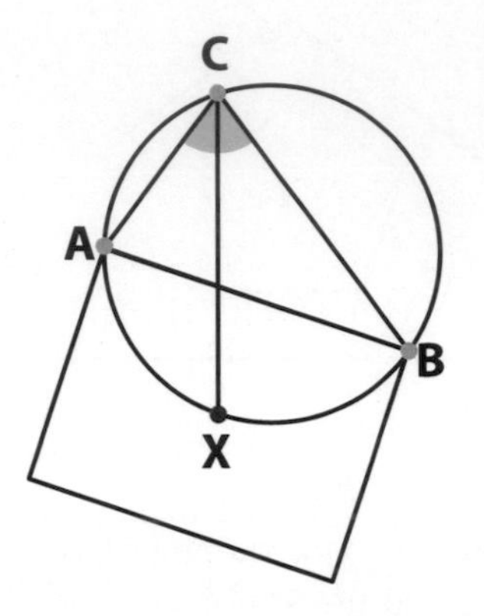

We want to show that X is at the center of the picture. We know that: 1) the center of the picture is on the angle bisector; and 2) the center of the picture is on the perpendicular bisector of AB. So, if we can show that X is also on the perpendicular bisector of AB, that will imply that X is at the center of the picture. To do this, we just have to show that AX and BX are of equal length, but this follows from the subtended green angles ACX and XCB being equal.

Did the last statement confuse you? Well, remember that all angles inscribed in a circle and subtended by the same chord are equal, and that different cord lengths correspond to different angles.

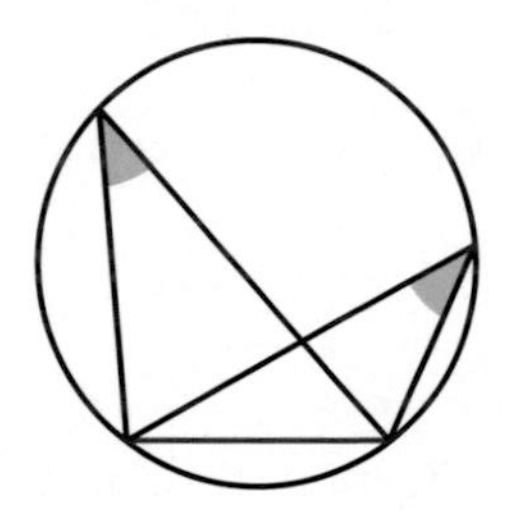

14. Tractrix and truck tricks.

The gradient of the truck equals the gradient of the tractrix at the back wheel. This gives us the differential equation

$$\frac{\mathrm{d}y}{\mathrm{d}x} = \frac{-\sqrt{L^2 - x^2}}{x}.$$

This equation can then be solved using standard calculus tricks, to give

$$y = \log_e \left(\frac{L + \sqrt{L^2 - x^2}}{x} \right) - \sqrt{L^2 - x^2}.$$

15. Cycling in circles.

After reading this Chapter it should be clear that the two rings have exactly the same area.

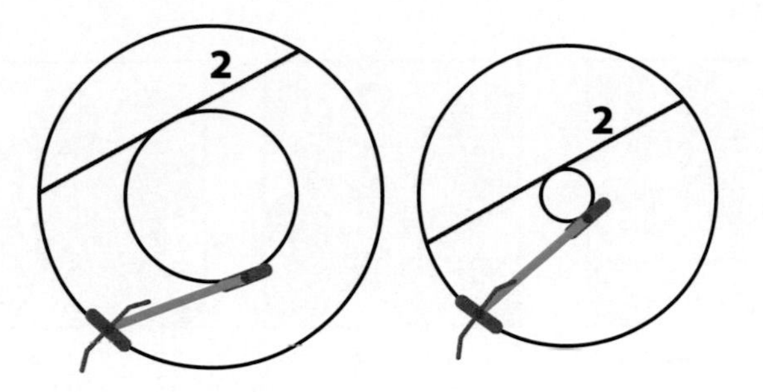

16. Which way did Natalie go?

Natalie traveled from left to right.

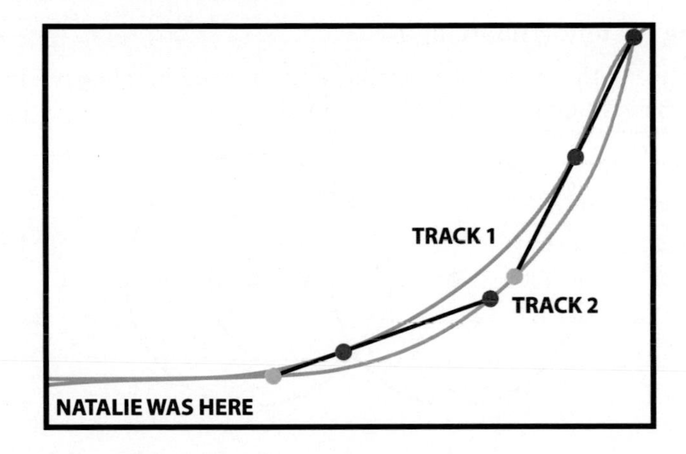

17. $\pi = 3$.

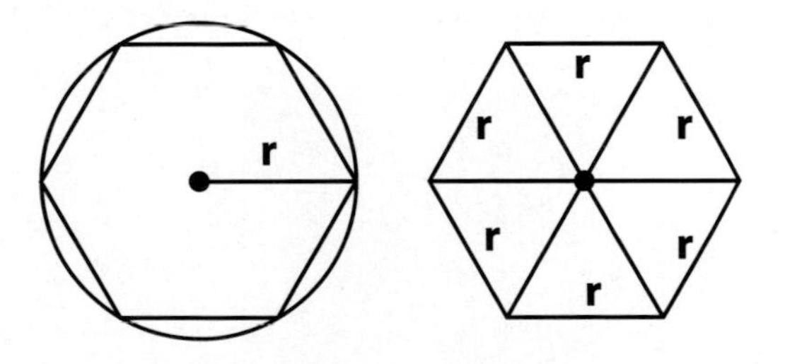

18. Just the right level of wine.

It turns out that one of our stacks will behave nicely exactly if none of the orange circles in the second row crosses the yellow line through the centres of the circles in the bottom row.

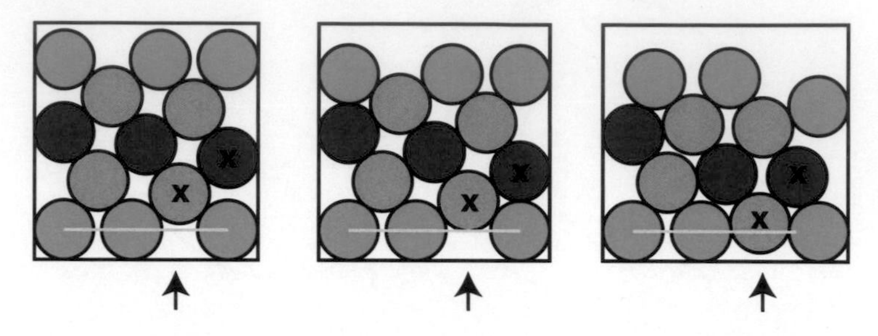

Starting with the diagram on the left, you can see what happens when we open up the gap that the arrow is pointing at. Pay particular attention to how the circles marked with an x move as the gap opens up.

19. Spotting an unfortunate spot.

With the centres of the three seismographs all contained in a line it's impossible to tell whether the earthquake occurred at the red or at the green location.

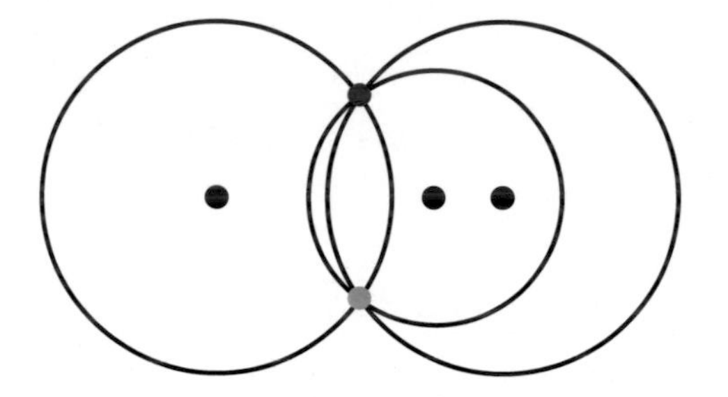

21. Choc-full of mathematics.

If the radius of the Mozartkugel is R, then the equator of this Mozartkugel has length $2\pi R$. This would also be the length of the green diagonal.

Then, if r is the radius of the arcs we are after, we calculate

$$(2r)^2 + (2r)^2 = (2\pi R)^2 ,$$

and so $r = \frac{\pi R}{\sqrt{2}}$.

22. Too hot, too cold, just right.

Let G stand for the temperature where Goldilocks happens to be. Similarly, let B stand for the Bear's temperature. Of course, as the two protagonists race around the house, G and B will continue to change. Now, at some time G will be at the hottest point of the circle, which means that at that moment $G > B$. (Unless, of course, there is more than one hottest point, and the Bear is at another. But in that case Goldilocks and the Bear are at the same temperature, which is the scenario we were hunting for.) Then, at some later time, the Bear will be at the hottest point, when we'll have $G < B$. So, by IVT, there is some middle time when $G = B$.

If this seems a little sleight of hand, you can make it less so by defining $D = G - B$, to be the difference in temperatures. Then, you can apply IVT to D.

23. Seeds of doubt.

First, suppose we are sticking to grand slam rules. Then there are 2 essentially different ways of arranging the four top seeds into the two halves of the draw: 1 and 4 are in the same half, or 1 and 3 are in the same half. With the top four seeds in place, we have to figure out how many ways there are to assign seeds 5-8 to the four quarters of the draw. But that's easy: there are 4 spots we can place the 5th seed, and then 3 spots left to place the 6th seed, 2 spots for the 7th seed, and finally 1 spot (i.e. no choice) for the 8th seed. Multiplying, we have a total of $2 \times 4 \times 3 \times 2 \times 1 = 48$ ways to arrange the top eight seeds.

Counting all possible seedings is similar, but trickier. First, let's see how many different ways there are of pairing the eight seeds into four pairs. Taking one player, there are 7 choices of whom to pair in the same quarter. That leaves six players, and taking one of those players, there are 5 choices of who to pair them with. That leaves four players, take one of them and there are 3 choices, and then finally 1 choice (i.e. no choice) for the final pair. We now have four pairs of players, one pair for each quarter, and we have to count the number of ways to arrange those quarters into two halves. It's easy to see there are 3 possible ways to do that. So, in total there are $7 \times 5 \times 3 \times 1 \times 3 = 315$ ways to arrange the top eight seeds.

24. Tennis math, anyone?

Here's a somewhat contrived tournament to show how to deal with negative multiples. Suppose we start with 64 players in a tennis tournament. Then in the first round there should be 32 matches. Imagine, however, that in 16 of those matches *both* players are injured. Further, imagine that these cancelled matches are spaced evenly throughout the draw, so that all the second round matches are walkovers. So, for each match played in the first two rounds, three players are eliminated. Continuing with the same pattern, and considering the total number of matches played in the six rounds, in two different ways, we have

$$32 - 16 + 8 - 4 + 2 - 1 = \frac{64 - 1}{3}.$$

25. The ball was in AND out? You cannot be serious!

The simplest way to fix the problem is to define it away. So, we'll declare a ball to be out if one of the linesmen calls "Out!", and otherwise the ball is in.

26. Giving it your best shot.

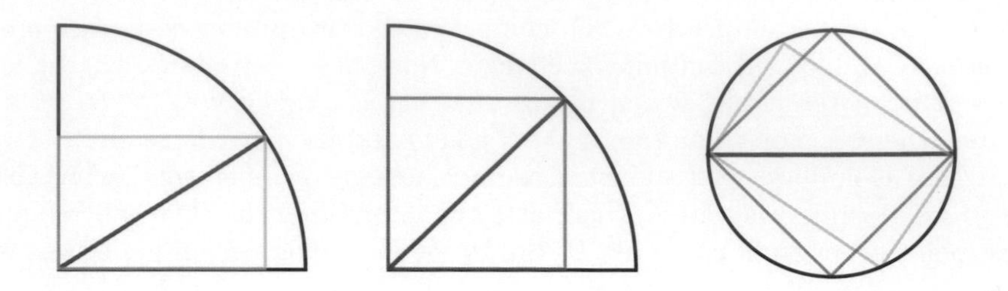

27. Bombs, and a bombed Riewoldt.

There are $4^4 = 256$ ways that four bombs might drop on the four squares, and $3^4 = 81$ of these ways avoid our given square. So, the chances are $81/256 \approx 32\%$ that our square will not be hit. Since the average bombs per square is 1, our Poisson formula estimates this probability to be $e^{-1} \approx 37\%$.

With 16 bombs on 16 squares, the average bombs per square is still 1, and so Poisson gives the same estimate of a 37% chance of no bombs on a given square. The exact probability of no bombs is $(15/16)^{16} \approx 36\%$. So, in this case the Poisson estimate is much closer.

The explanation for all this comes down to the definition of e: see Chapter 7. Specifically, just as e is the limit of $(1 + 1/N)^N$ as N gets huge, it turns out that $1/e$ is the limit of $(1 - 1/N)^N = (\frac{N-1}{N})^N$. This expression is exactly our no-square probability, and so the more squares (with the identical number of bombs) we use, the closer our probability will be to $1/e$.

28. Diophantine footy fan.

Suppose G is a whole number and N is a common factor of both G and $G - 1$. Then for suitable whole numbers A and B, we can write $G = AN$ and $G-1 = BN$. Equating the two expressions for G, this gives $AN = 1+BN$, and rearranging gives $(A-B)N = 1$. Since we have a product of whole numbers, that is impossible unless $N = 1$ (or also $N = -1$, if we're also allowing negatives).

If a goal is worth 5 points, then our "product equals total" equation is $5G+B = GB$. Rearranging, this gives $5G = (G - 1)B$. It follows that $G - 1$ is a factor of 5, leading to the two solutions, $G = 2, B = 10$ and $G = 6, B = 6$. And, there is the "St. Kilda" solution, $G = 0, B = 0$.

29. Walk, don't run!

Suppose, as suggested, Jared is able to lift both feet off the ground for 1/25 of a second without it being detected. If Jared is walking at 15 kilometers per hour, then he can progress about 17 centimeters in that time.

30. Tour de math

A silly way to solve this puzzle is to simply have a single 10-tooth back sprocket together with the appropriate front sprockets. A more reasonable solution is given by the second table.

	10
12	1.2
15	1.5
16	1.6
20	2.0
24	2.4
25	2.5
30	3.0

	4	5
6	1.5	1.2
8	2.0	1.6
10	2.5	2.0
12	3.0	2.4

Of course, if you're not happy with four teeth on a sprocket, you can always double or triple the numbers of teeth on all the sprockets, etc.

31. How round is your soccer ball?

The shape below is called a *Steinmetz solid*. It is the intersection of three identical and mutually perpendicular cylinders whose axes meet at a point. As such it has three circular shadows.

32. And the winner is ...

	short program	before Irina		after Irina	
		long program	total	long program	total
Michelle Kwan	0.5	3.0	3.5	4.0	4.5
Sarah Hughes	3.0	1.0	4.0	1.0	4.0
Sasha Cohen	2.0	2.0	4.0	2.0	4.0
Irina Slutskaya	1.0			3.0	4.0

33. Visionary Voronoi . . .

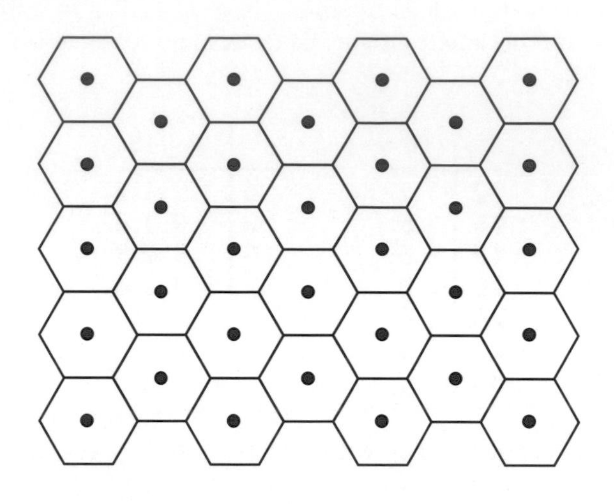

34. Melbourne's catenary chaos

Flipping things upside down we get this:

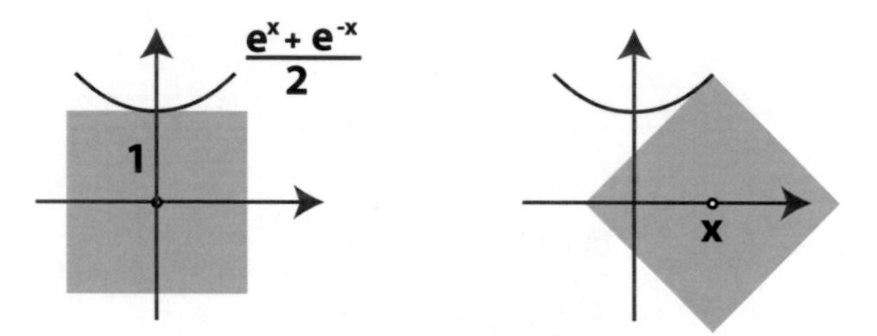

The catenary wavelength we're after is twice the x in the picture on the right. The vertex of the square is at height $\sqrt{2}$. Therefore we have

$$\frac{e^x + e^{-x}}{2} = \sqrt{2}\,.$$

This is a quadratic equation in e^x with positive solution $1 + \sqrt{2}$. Therefore

$$x = \log_e(1 + \sqrt{2}) \approx 0.88\,,$$

and so the wavelength we're after is

$$2\log_e(1 + \sqrt{2}) \approx 1.76\,.$$

35. Eureka!

Reading off the diagram we have the three different circle areas:

(1) circle in cylinder $= 4\pi R^2$;
(2) circle in cone $= \pi H^2$;
(3) circle in sphere (using Pythagoras) $= \pi(2RH - H^2)$.

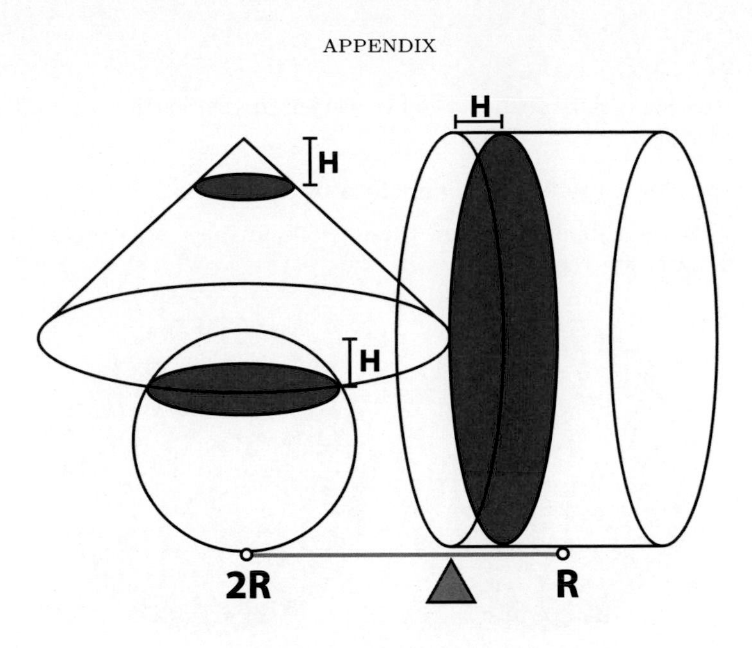

The weight on the left multiplied by the distance to the fulcrum is

$$(\pi H^2 + \pi(2RH - H^2)) \times 2R = 4\pi R^2 H\,.$$

The weight on the right multiplied by the distance to the fulcrum is

$$4\pi R^2 \times H = 4\pi R^2 H\,.$$

Since these two products are equal the law of the lever guarantees that the two circles on the left balance the circle on the right.

36. Archimedes' crocodile.

Stretching a plane figure with area A by a factor R in the direction of the x- or y-axis gives a new figure of area AR. To turn a unit circle of area π into a circle of radius R we stretch in the directions of both the x- and the y-axes. This means that this new circle has area πR^2.

37. Melbourne Grammar mystery map.

If the bear lives in the northern hemisphere, then the information provided implies that the bear's house is right at the North Pole, and the bear is white.

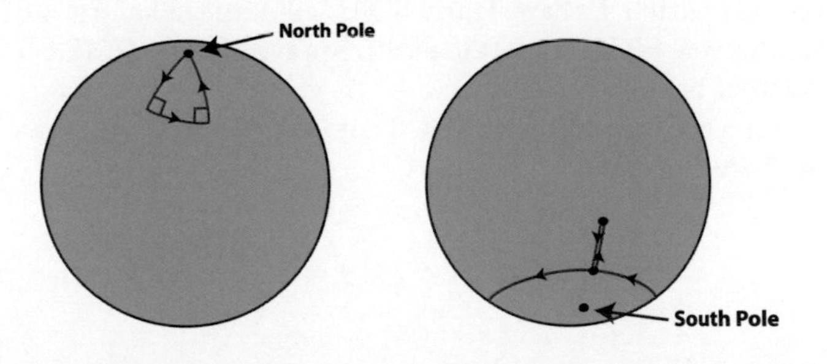

If the bear were a bird then it could be very near the South Pole, and he would be black and white.

38. A rectangle, some spheres and lots of triangles.

There are as many rubber bands on these geodesic spheres as there are pairs of diametrically opposite vertices.

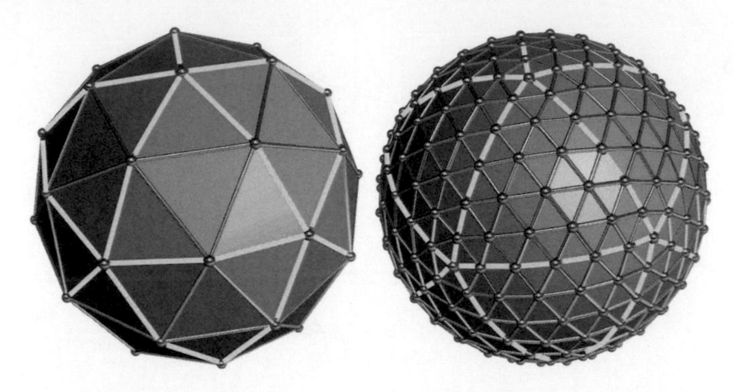

The underlying yellow icosahedron has 12 vertices, 20 faces and 30 edges. The vertices of the geodesic sphere on the left are these 12 original vertices plus an extra vertex per edge of the icosahedron. This means the geodesic sphere on the left has $12 + 30 = 42$ vertices, and therefore the corresponding rubber band ball has $42/2 = 21$ rubber bands. Similarly, counting up the vertices on the right geodesic sphere, it follows that the other rubber band ball consists of

$$(12 + (5 \times 30) + (20 \times 10))/2 = 181$$

rubber bands.

39. So you think you can beat the bookies.

Betting $\$1/A$, $\$1/B$ and $\$1/C$, respectively, on the three horses guarantees a return of \$1. If the first horse wins, for example, then your winnings will $A \times 1/A = \$1$.

Once Son of Seabiscuit has entered the race, you'll have to bet an extra \$1/100, that is, one extra cent, to guarantee your \$1 return. So, the rigging has gone up, albeit by a minuscule amount. This example illustrates a very important point about gambling: the fact that the *overall* odds on a race are rigged does not preclude the possibility that some *specific* bet on the race is to the bettor's advantage.

40. A Penney for your thoughts.

There are four equally likely ways that the coin-tossing could begin: TT, TH, HT and HH. If we start with TT then clearly TTH will appear before HTT. Once an H has occurred, however, HTT is guaranteed to appear before TTH. So, overall HTT has a 3/4 chance of beating TTH.

Just by symmetry, swapping H and T, it is clear that HTT and THH must have an equal chance of winning.

41. The Playmobil mystery.

First imagining we buy the cards one at a time, on average to get the 72 different cards, the number of cards we will need to purchase is

$$1 + \frac{72}{71} + \frac{72}{70} + \cdots + \frac{72}{3} + \frac{72}{2} + \frac{72}{1}.$$

That comes to about 350 cards, and so we'll have to purchase about 117 packs of three cards.[13]

For people who aren't thrilled about summing seventy-two fractions, there is an alternative, calculus approach. To begin, it is natural write the leading 1 in the above sum as $\frac{72}{72}$, and factor out the 72. Then we can write the sum as

$$72\left(\frac{1}{1} + \frac{1}{2} + \frac{1}{3} + \cdots + \frac{1}{70} + \frac{1}{71} + \frac{1}{72}\right).$$

The stuff inside the brackets is part of the famous *harmonic series*: see Chapter 4. A very approximate answer to the bracket sum is the integral

$$\int_1^{72} \frac{1}{x}\,\mathrm{d}x = \log_e 72 \approx 4.28\,.$$

Then, to make the approximation much more accurate, we can add in the *Euler-Mascheroni constant*, $\gamma \approx 0.58$.[14] That gives the bracket sum to be about 4.86, and multiplying by 72 again gives us our 350 cards to purchase.

Finally, it is also worth noting that using a geometric series trick – see Chapter 24 – the bracket sum can be written exactly as

$$\int_0^1 \frac{1-x^{72}}{1-x}\,\mathrm{d}x\,.$$

42. The devil is in the dice.

The tail of dice is probably fake since, among other things, no two consecutive dice show the same number.

The average rolls required to obtain two 6s in a row can be calculated in the same manner as for one 6, but is a little trickier. Again, we'll let A stand for the average number of rolls. We'll also let S stand for the average number of extra rolls required *after* we've rolled a 6.

Now think about the situation after one roll. We have a $5/6$ chance of *not* rolling a 6, and being back at square one. But we also have a $1/6$ chance of rolling

[13] And did purchase. We're still trying to get rid of the gum.

[14] A story for another book.

a 6, and then the average number of extra rolls from there is S. So, in total A must satisfy the equation.

$$A = 1 + \left(\frac{5}{6} \times A\right) + \left(\frac{1}{6} \times S\right).$$

Now, let's see what we can figure out about S. We imagine we've rolled a 6, and now, where can we be after one more roll? Well, if we roll another 6 then we're done, but there's a 5/6 chance of not rolling a 6, bringing us back to square one. That tells us that S must satisfy the equation

$$S = 1 + \left(\frac{5}{6} \times A\right).$$

And, now we're home. Substituting the expression for S into the equation for A, it easily comes out that $A = 42$ (and $S = 36$).

43. The Freddo Frog path to perfection.

If Bert found B errors and Ernie found C of Bert's errors, then Ernie found C/B of Bert's errors. So, if we assume that Ernie found the same fraction C/B of the total errors, and if Ernie found E errors, then our estimate for the total is $E/(C/B) = EB/C$. By symmetry, thinking of Bert finding Ernie's errors gives the same estimate.

44. Will Rogers, clever Kiwis and medical magic.

If we don't want to lower the average of Group B, then the students we transfer to group A must have an average no greater than 70. Since the average of 30 and 70 is 50, this shows that we can't raise the average of Group A above 50 without also lowering the Group B average. Now, imagine one student in Group B scored 100, and transfer everyone but this genius into Group A. Then the average of Group B will be 100, and the average of group A will be 49.5.

45. The hidden karma of Snakes and Ladders.

We simply have to consider the different ways of getting to 10 on one or two rolls. For example, if we roll a 1 on our first go, then the only chance to get to 10 on the next turn is to roll another 1: there's a $1/6 \times 1/6 = 1/36$ chance of that. Similarly, there's a $1/36$ chance if we roll a 2 first up. Next, there's a $1/6$ chance of rolling a 3, and so getting to 10 in a single move, and so on. Summing up all the possibilities gives a $5/36 + 1/6 = 11/36$ chance of escaping in one or two moves.

46. Poet of the Universe.

The "primum mobile" would be the equator (or any great circle), and the two worlds would be the hemispheres on either side.

47. Escape to our Moon planet.

Let's mark a point on the perimeter of the rolling coin/circle, start the circle rolling, and track the marked point.

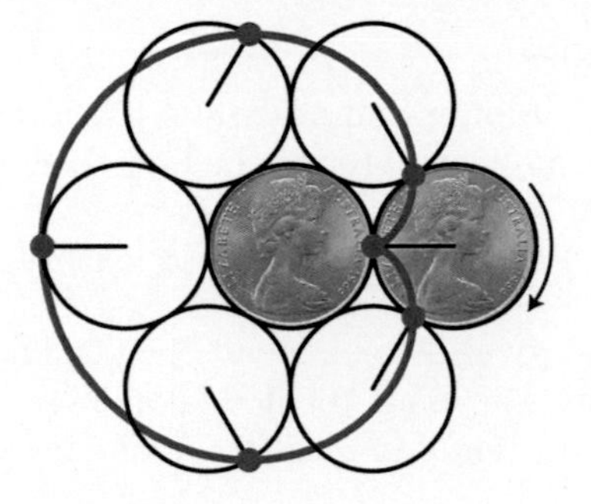

As the diagram indicates, the rolling coin will rotate twice.

48. Tickling Orion with a triangle.

Take an equilateral triangle on the flat surface. Adding the vertical lines (sun rays) through the vertices we get a right equilateral triangular prism. Therefore what we have to show is that given any green triangle there is such a prism whose three long edges contain the vertices of the triangle. The idea for the proof is very similar to that presented in the Chapter.

To see that this can always be arranged, start by embedding the triangle in a prism with small enough cross-section, as pictured.

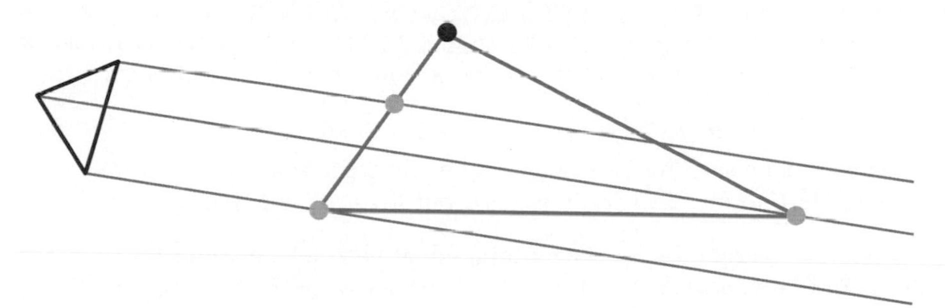

Now expand the cross-section of our prism such that at all times the three orange points remain on the edges of the prism. In this way, eventually all three vertices of the triangle will be contained in the edges of the prism.

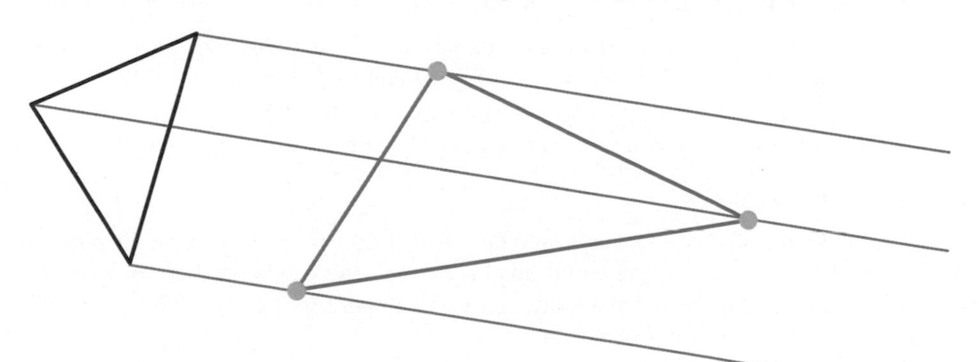

49. The eternal triangles.

The centre of mass of the Jupiter-Sun system is the lever point. So, if the center of mass is D kilometers from the center of the Sun then we must have

$$\left(2 \times 10^{30}\right) D = \left(2 \times 10^{27}\right) \left(8 \times 10^{8} - D\right) .$$

Cancelling things out, that gives D to be about 800,000 kilometers. That is interesting because the Sun's radius is about 700,000 kilometers. So, Jupiter is sufficiently massive for the Jupiter-Sun center of mass to lie outside the Sun.

50. On primes and Pluto.

First, notice that $(3+2\sqrt{2})(3-2\sqrt{2}) = 1$. What this tells you is that any "factorisation" of a number in our world can also include as many factors $(3+2\sqrt{2})(3-2\sqrt{2})$ as you like, just as multiple 1s can be included. Next, notice that $7 = (3+\sqrt{2})(3-\sqrt{2})$, so definitely 7 seems to have a weird factorisation in our world.

Such worlds are exactly why mathematicians started thinking seriously about "primes" and "ones". In fact, our $a + b\sqrt{2}$ world contains infinitely many "ones", referred to as *units* by the professionals. Indeed, it is not immediately obvious that 7 and/or $3 \pm \sqrt{2}$ are not units, although they are not, and this is pretty easy to prove. As it happens, and this is *not* so easy to prove, the fundamental theorem of arithmetic holds in our $a+b\sqrt{2}$ world: every number in this world can be factorised as a product of primes (numbers that cannot be broken down further), and this factorisation is unique except for the ordering and for the choice of units.

51. The math of planet Mars.

Numbers are numbers. So, for example we might write the number thirteen as 13, while our Martian friend writes it as 1101, but it's the same number.

These are tricky questions, depending upon how numbers and arithmetic are defined. In brief, however, $1 + 1 = 2$ is true by the definition of "2"; it is the next number, or thing, after 1. On the other hand, 4 is naturally defined as the next number after 3, so as $3 + 1$. Then, the fact that $2 + 2 = 3 + 1$ is a consequence of the way we define addition.

52. The eternal grind.

We'll do a rough calculation, using the fact that $2^{10} \approx 1000$. So, each ten folds multiplies the thickness by 1000. Then twenty folds multiplies by a million, thirty folds by a billion, and so on. We want to multiply by $10 \times 1000 \times 150,000,000$. So, forty folds will get us about 2/3 of the way to the sun. Then, one more fold...

53. Calendar kinks.

If we want equal, 30-day months then the natural approach is to have 12 months, plus five QED Cat Days. One could have a special day at the end of each season, and one more at the end of the year. Also, God permitting, the 30-day months would naturally fit with 6-day weeks.

54. Strange moves of a mathematical feast.

The formula calculates the day and month of Easter in the year Y. Here $\lfloor x \rfloor$ denotes the integer part of the number x, and x_n denotes the remainder of x after division by n.

55. Lucky Friday the 13th.

There's nothing to do but go through it. In the 400-year cycle of the Gregorian Calendar, the 13th is a Friday 56 times in August and October, 57 times in February, June, September and December, and 58 times in each of the remaining six months.

56. Hermann the hermit.

Write the answer to your puzzle below:

57. The equals of Robert Recorde.

if you abate even portions from thynges that bee equalle, the partes that remain shall be equall also.

Of course $-1 = -1$, and $2/6 = 1/3$. Therefore, it must be true that $(-1)^{\frac{2}{6}} = (-1)^{\frac{1}{3}}$.

58. Pythagoras's theorem ain't Pythagoras's.

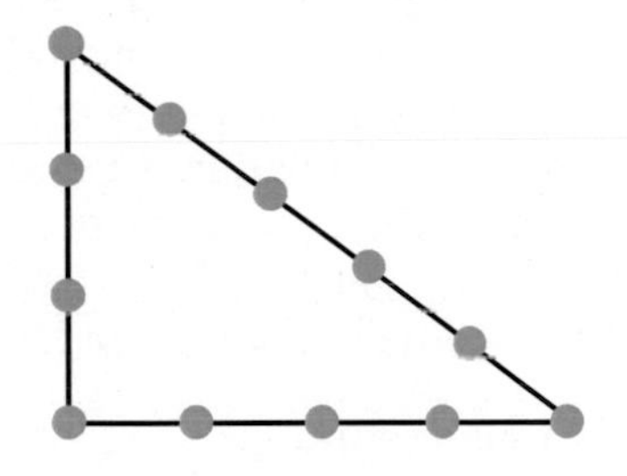

59. Six of one, Babylonian the other.

$$\pi \approx 3.8 = 3 + \frac{8}{60}.$$

$$\frac{1}{7} = 0.\overline{(8)(34)(17)}.$$

60. It's Chris Mass time.

11 minutes and 14 seconds per year, for 400 years, comes to 3 days 2 hours and 53 minutes. The Gregorian calendar eliminates 3 leap days each 400 years, and so will be about three hours behind after 400 years. Our current Time Gods take care of this by eliminating the occasional "leap second".

61. Clearing a logjam.

We have

$$\begin{aligned}\text{nlog}(ab) &= -10^7\left(\log_e(ab) - \log_e\left(10^7\right)\right)\\ &= -10^7\left(\log_e a + \log_e b - \log_e\left(10^7\right)\right)\\ &= -10^7\left(\log_e a - \log_e\left(10^7\right)\right) - 10^7\left(\log_e b - \log_e\left(10^7\right)\right) - 10^7\log_e\left(10^7\right)\\ &= \text{nlog}(a) + \text{nlog}(b) - 10^7\log_e\left(10^7\right).\end{aligned}$$

62. Squares, triangles and other labor-saving devices.

Since $T(N) = N(N+1)/2$, we have

$$2T(A-1)+2T(B)-2T(A-B-1) = (A-1)A+B(B+1)-(A-B-1)(A-B).$$

It is easy to check that everything on the right cancels except for $2AB$.

63. The doodle, the witch and Maria.

The circle with center $(0, \frac{1}{2})$ and radius $\frac{1}{2}$ has equation $x^2 + \left(y - \frac{1}{2}\right)^2 = \frac{1}{4}$, which, after expanding the y term, simplifies to

$$x^2 + y^2 - y = 0.$$

Now, write the equation of the straight line leaving the origin as $x = ky$. This line will hit the horizontal line $y = 1$ when $x = k$, which gives us the x-coordinate of our witch-point. As well, substituting the line into our circle equation, we obtain $k^2y^2 + y^2 - y = 0$. One solution to this equation is $y = 0$, which corresponds to the origin. Ignoring this solution and dividing by y, we obtain the solution we're after: $y = 1/(1+k^2)$. In summary, the coordinates of our witch-point are $(k, 1/(1+k^2))$. In other words, for any point (x, y) on our witch, we have $y = 1/(1+x^2)$.

64. Christian Goldbach's magic sum.

$4 = 2+2, 6 = 3+3, 8 = 3+5, 12 = 5+7$, and that's it. All other even numbers greater than 2 and below 100 can be written as a sum of two primes in at least two different ways.